SpringerBriefs in Applied Sciences and Technology

Safety Management

Series editors

Eric Marsden, FonCSI, Toulouse, France
Caroline Kamaté, FonCSI, Toulouse, France
François Daniellou, FonCSI, Toulouse, France

The SpringerBriefs in Safety Management present cutting-edge research results on the management of technological risks and decision-making in high-stakes settings.

Decision-making in high-hazard environments is often affected by uncertainty and ambiguity; it is characterized by trade-offs between multiple, competing objectives. Managers and regulators need conceptual tools to help them develop risk management strategies, establish appropriate compromises and justify their decisions in such ambiguous settings. This series weaves together insights from multiple scientific disciplines that shed light on these problems, including organization studies, psychology, sociology, economics, law and engineering. It explores novel topics related to safety management, anticipating operational challenges in high-hazard industries and the societal concerns associated with these activities.

These publications are by and for academics and practitioners (industry, regulators) in safety management and risk research. Relevant industry sectors include nuclear, offshore oil and gas, chemicals processing, aviation, railways, construction and healthcare. Some emphasis is placed on explaining concepts to a non-specialized audience, and the shorter format ensures a concentrated approach to the topics treated.

The SpringerBriefs in Safety Management series is coordinated by the Foundation for an Industrial Safety Culture (FonCSI), a public-interest research foundation based in Toulouse, France. The FonCSI funds research on industrial safety and the management of technological risks, identifies and highlights new ideas and innovative practices, and disseminates research results to all interested parties.

For more information: https://www.foncsi.org/.

FONCSI

Fondation pour une culture
de sécurité industrielle

More information about this series at http://www.springer.com/series/15119

Siri Wiig · Babette Fahlbruch
Editors

Exploring Resilience

A Scientific Journey from Practice to Theory

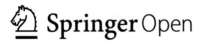

Editors
Siri Wiig
Faculty of Health Sciences, SHARE-Centre
for Resilience in Healthcare
University of Stavanger
Stavanger, Norway

Babette Fahlbruch
TÜV Nord EnSys GmbH & Co. KG
Berlin, Germany

ISSN 2191-530X   ISSN 2191-5318   (electronic)
SpringerBriefs in Applied Sciences and Technology
ISSN 2520-8004   ISSN 2520-8012   (electronic)
SpringerBriefs in Safety Management
ISBN 978-3-030-03188-6   ISBN 978-3-030-03189-3   (eBook)
https://doi.org/10.1007/978-3-030-03189-3

Library of Congress Control Number: 2018959268

This Springer imprint is published by the registered company Springer Nature Switzerland AG
The registered company address is: Gewerbestrasse 11, 6330 Cham, Switzerland

Contents

Chapter 1
Exploring Resilience – An Introduction

Siri Wiig and Babette Fahlbruch

Abstract Resilience has become an important topic on the safety research agenda and in organizational practice. In this chapter we give an introduction to the research area and some of the current challenges, before we present the aim of the book.

Keywords Resilience · Safety · Research · Organizational practice · Theoretical framework

1.1 Resilience – What Is It?

Resilience has become an important topic on the safety research agenda and in organizational practice (e.g. [1–6]). Numerous definitions of resilience exist within different research traditions, disciplines, and fields such as sociology, psychology, medicine, engineering, economics, ecology, political science [7–11]. The common use of the resilience concept relates to the ability of an entity, individuals, community, or system to return to normal condition or functioning after the occurrence of an event that disturbs its state.

Many similarities can be observed across the resilience concept applications [8, 9, 12]. We often see resilience research literature referring to dynamic capabilities, adaptive capacity, and performance variation as key topics. Some group resilience literature into three general areas related to readiness and preparedness; response and adaptation; and recovery or adjustment and argue that researchers attempt to broadly cover all three areas in one study, but individually each area receives limited attention resulting in a diverse literature base [9]. Others have identified domains of

S. Wiig (✉)
Faculty of Health Sciences, SHARE-Centre for Resilience in Healthcare,
University of Stavanger, Stavanger, Norway
e-mail: siri.wiig@uis.no

B. Fahlbruch
TÜV NORD EnSys, Berlin, Germany
e-mail: bfahlbruch@tuev-nord.de

S. Wiig and B. Fahlbruch (eds.), *Exploring Resilience*, SpringerBriefs
in Safety Management, https://doi.org/10.1007/978-3-030-03189-3_1

resilience such as: the Organizational domain – addressing the need for enterprises to respond to a rapid changing business environment; the Social domain – addressing capabilities of individuals, groups, community and environment to cope with external stress; the Economic domain – addressing the inherent ability and adaptive response that enable firms and regions to avoid maximum potential loss; and the Engineering domain – which is mainly adopted within in the safety science as the intrinsic ability of a system to adjust its functionality in the presence of disturbance and unpredicted changes [8]. The resilience engineering (e.g. [13, 14]) domain has attracted a wide readership in the safety science the past decade, although it has a history going back to the 1980s and rooted in cognitive system engineering, human factors, and system safety engineering [3].

1.2 Some Current Challenges

The shared use of the resilience term across different traditions does not imply unified concepts of resilience nor theories in which it is embedded [12]. In the resilience literature in more general there has been a strong focus on building theories, however there is lack in empirically proving the theories [9, 15]. This is also true for resilience as it is used in the safety science [10, 14]. The current body of knowledge on complex adaptive systems and resilience has increased our understanding of organizations and the challenges they face in particularly in relation to social and technological complexity, but it suffers from being too generalized and abstract. Identification of what constitutes resilience has hardly been clarified under the onslaught of theorizing and individual empirical cases [16]. A recent systematic review demonstrates that some scientific efforts have been made to develop constructs and models that present relationships; however, these cannot be characterized as sufficient for theory building [10, 17]. Other attempts to model resilience theoretical frameworks (e.g. [18]) lack empirical testing. The current lack of well-defined constructs is a scientific drawback within the safety science, as it is too unclear which phenomena are to be operationalized [10].

There is a need to develop a coherent integrative theoretical framework of resilience mechanisms to enable large-scale comparative longitudinal studies across multiple high-risk settings and sectors (e.g. healthcare, transport, petroleum, nuclear power) and countries [19, 20]. A major current research challenge is the absent integration of different system levels from individuals, teams, organizations, regulatory bodies, and policy level [10, 14, 21, 22], implying that mechanisms through which resilience is linked across the micro/meso/macro level are not yet well understood. For example, most current research addresses activities of front-line workers (micro level) (e.g. [1]) and stresses factors of work system design, while top management teams (meso level) [23] and leadership for organizational adaptability [24], external contextual factor and regulatory system (macro level) are lacking as key resilience dimensions in theoretical frameworks. Regulation is often the first lever that policy makers and professional bodies reach for to drive improvements in safety, yet the relationship between regulation and resilience remains little explored and the role of regulation in producing or potentially undermining resilience performance, needs investigation and theorizing [21, 25].

The role of stakeholders in resilience is underexplored. Despite the literature within for example healthcare focusing on patient and next of kin as co-creators of resilience, studies lack involvement of stakeholders [26, 27]. High-risk industries depend on collaboration across numerous stakeholders, of potential influence on resilience within organizations and in a societal perspective. In order to understand how individuals, groups, organizations and communities need to adapt and respond to internal and external change and context, stakeholder analysis (e.g. [28]) could add to the body of knowledge in resilience. This is also of relevance for the practical and operational approaches to resilience in terms of developing targeted strategies for different stakeholders and to establish for example collaboratives for sharing knowledge across levels to foster resilience when it depends on inter-professional collaboration and collaboration across system interfaces (e.g. [29, 30]), and across different conceptualizations of resilience, safety and security which are often in contradiction (e.g. [31, 32]). Currently, this area needs exploration of new approaches to ensure operationalization of resilience as a multi-stakeholder phenomenon.

The latter illustrates that there are not only theoretical and empirical research challenges related to resilience research. There are also challenges related to the translation of theory into practice by providing practical guidance to different stakeholders, on how to design and operate resilient organizations and to maintain resilience (e.g. [33]). There is a need for developing testable propositions and interventions related to resilience and exploring this in guided iterative cycles of design and evaluation [10, 11]. However, how this best should be operationalized is still unclear. We argue in line with [1] that it would be interesting to widen the perspective of resilience applied in the safety domain by looking at how other scientific domains operationalize it, and through this may gain new insight and possible improvement in both theory building and translation of theory into interventions and practical solutions.

1.3   What Is This Book Looking for?

This book does not advocate for one definition or one field of research when talking about resilience; it does not assume that the use of resilience concepts is necessarily positive for safety. We encourage a broad approach, seeking inspiration across different scientific and practical domains for the purpose of further developing resilience at a theoretical and an operational level of relevance for different high-risk industries. The aim of the book is twofold:

1. To explore different approaches for operationalization of resilience across scientific disciplines and system levels.
2. To create a theoretical foundation for a resilience framework across scientific disciplines and system levels.

By presenting chapters from leading international authors representing different research disciplines and practical fields we develop suggestions and inspiration for the research community and for practitioners in high-risk industries.

References

1. J. Bergström, R. van Winsen, E. Henriqson, On the rationale of resilience in the domain of safety: a literature review. Reliab. Eng. Syst. Saf. **141**, 131–141 (2015)
2. C.P. Nemeth, I. Herrera, Building change: resilience engineering after ten years. Reliab. Eng. Syst. Saf. **141**, 1–4 (2015)
3. J.-C. Le Coze, Vive la diversité! High reliability organisation (HRO) and resilience engineering (RE). Saf. Sci. (2016, In press)
4. E. Hollnagel, J. Braithwaite, R.L. Wears (eds.), *Resilient Health Care* (Ashgate, Farnham, 2013)
5. J. Braithwaite, R.L. Wears, E. Hollnagel, Resilient health care: turning patient safety on its head. Int. J. Qual. Health Care **27**(5), 418–420 (2015)
6. M. Pillay, Resilience engineering: an integrative review of fundamental concepts and directions for future research in safety management. Open J. Saf. Sci. Technol. **7**(4), 129–160 (2017)
7. X. Xue, L. Wang, R.J. Yang, Exploring the science of resilience: critical review and bibliometric analysis. Nat. Hazards **90**(1), 477–510 (2018)
8. S. Hosseini, K. Barker, J.E. Ramirez-Marquez, A review of definitions and measures of system resilience. Reliab. Eng. Syst. Saf. **145**, 47–61 (2016)
9. R. Bhamra, S. Dani, K. Burnard, Resilience: the concept, a literature review and future directions. Int. J. Prod. Res. **49**(18), 5375–5393 (2011)
10. A.W. Righi, T.A. Saurin, P. Wachs, A systematic literature review of resilience engineering: research areas and a research agenda proposal. Reliab. Eng. Syst. Saf. **141**, 142–152 (2015)
11. A. Annarelli, F. Nonino, Strategic and operational management of organizational resilience: current state of research and future directions. Omega **62**, 1–18 (2016)
12. P. Martin-Breen, J.M. Anderies, Resilience: a literature review. Technical Report, The Bellagio Initiative (2011)
13. E. Hollnagel, D.D. Woods, N. Leveson (eds.), *Resilience Engineering: Concepts and Precepts* (Ashgate, Aldershot, 2006)
14. R. Patriarca, J. Bergström, G.D. Gravio, F. Costantinoa, Resilience engineering: current status of the research and future challenges. Saf. Sci. **102**, 79–100 (2018)
15. E.A.M. Limnios, T. Mazzarol, A. Ghadouani, S. Schilizzi, The resilience architecture framework: four organizational archetypes. Eur. Manag. J. **32**, 104–116 (2014)
16. K.A. Pettersen, P.R. Schulman, Drift, adaptation, resilience and reliability: toward an empirical clarification. Saf. Sci. (2016, in press)
17. R.I. Sutton, B.M. Staw, What theory is not. Adm. Sci. Q. **40**(3), 371–384 (1995)
18. J. Lundberg, B.J.E. Johansson, Systemic resilience model. Reliab. Eng. Syst. Saf. **141**, 22–32 (2015)
19. L.K. Comfort, A. Boin, C.C. Demchak, Resilience revisited - an action agenda for managing extreme events, in *Designing Resilience: Preparing for Extreme Events* (University of Pittsburgh Press, Pittsburgh, 2010)
20. M.A. Hitt, P.W. Beamish, S.E. Jackson, J.E. Mathieu, Building theoretical and empirical bridges across levels: multilevel research in management. Acad. Manag. J. **50**(6), 1385–1399 (2007)
21. C. Macrae, Reconciling regulation and resilience in health care, in *Resilient Health Care* ed. by E. Hollnagel, J. Braithwaite, R.L. Wears (Ashgate, Farnham, 2013)
22. J. Bergström, S.W.A. Dekker, Bridging the macro and the micro by considering the meso: reflections on the fractal nature of resilience. Ecol. Soc. **19**(4) (2014)

23. A. Carmeli, Y. Friedman, A. Tishler, Cultivating a resilient top management team: the importance of relational connections and strategic decision comprehensiveness. Saf. Sci. **51**, 148–159 (2013)
24. M. Uhl-Bien, M. Arena, Leadership for organizational adaptability: a theoretical synthesis and integrative framework. Leadersh. Q. **29**(1), 89–104 (2018)
25. R. Bal, A. Stoopendaal, H. van de Bovenkamp, Resilience and patient safety: how can health care regulations contribute? Nederlands Tijdschrift voor Geneeskunde **159** (2015)
26. C.C. Schubert, R. Wears, R.J. Holden, G.S. Hunte, Patients as a source of resilience, in *Resilient Health Care, Volume 2: The Resilience of Everyday Clinical Work*, ed. by R.L. Wears, E. Hollnagel, J. Braithwaite (Ashgate, Farnham, 2015), pp. 207–225
27. C. Vincent, R. Amalberti, *Safer Healthcare* (Springer, Berlin, 2016)
28. R. Brugha, Z. Varvasovszky, Stakeholder analysis: a review. Health Policy Plan. **15**(3), 239–246 (2000)
29. M. Storm, I. Siemsen, K. Laugaland, D. Dyrstad, K. Aase, Quality in transitional care of the elderly: key challenges and relevant improvement measures. Int. J. Integr. Care **14** (2014)
30. K.A. Laugland, Transitional care of the elderly from a resilience perspective. Ph.D. thesis, University of Stavanger (2015)
31. K.A. Pettersen, T. Bjørnskau, Organizational contradictions between safety and security - perceived challenges and ways of integrating critical infrastructure protection in civil aviation. Saf. Sci. **71**, 167–177 (2015)
32. R. Østgaard Skotnes, Challenges for safety and security management of network companies due to increased use of ICT in the electric power supply sector. Ph.D. thesis, University of Stavanger (2015)
33. E. Lay, M. Branlat, Z. Woods, A practitioner's experience operationalizing resilience engineering. Reliab. Eng. Syst. Saf. **141**, 63–73 (2015)

Chapter 2
Resilience, Reliability, Safety: Multilevel Research Challenges

Jean-Christophe Le Coze

Abstract This chapter contributes to current research on resilience by considering two aspects of this topic. The first describes the popularity of resilience as a product of a shift of era which creates a degree of uncertainty about the future in several domains of concern in a globalised context, and how this notion has also travelled in the field of safety. The second part addresses the cognitive, institutional, methodological, empirical and theoretical challenges of interdisciplinary multilevel safety research.

Keywords Multilevel research · Methodology · Interdisciplinarity · Fieldwork Ethnography

2.1 A Resilience Moment

The success of the notion of resilience is to be understood in the context of a shift of era. Climate change related events, economic turmoil of countries and companies in international competition, technological developments with uncertain consequences, identities struggling under macro globalised processes shaking the status of nation-states... these trends have created a favourable background for terrorist attacks, natural catastrophes, technological disasters and financial crisis which have in turn created a very favourable moment in history for a notion such as resilience.

Globalisation processes of increased flows of people, goods, information, images, money across the space and time beyond anything comparable in previous epochs have created entire new contexts for nation-states, businesses and populations... and safety. New kind of threats from and to critical infrastructures of our societies — energy, transport, medical, administrative, informational networks — now exist. These threats are both endogenous in terms of managing their sheer complexity and exogenous in terms of terrorists or natural catastrophes exposures for instance. They are global [1, 2].

J.-C. Le Coze (✉)
Ineris, Verneuil-en-Halatte, France
e-mail: lecoze@ineris.fr

© The Author(s) 2019
S. Wiig and B. Fahlbruch (eds.), *Exploring Resilience*, SpringerBriefs
in Safety Management, https://doi.org/10.1007/978-3-030-03189-3_2

In this situation, resilience, defined as the ability *"to proactively adapt to and recover from disturbances that are perceived within the system to fall outside the range of normal and expected disturbances"* [3], or as *"the intrinsic ability of an organisation (system) to maintain or regain a dynamically stable state, which allows it to continue operations after a major mishap and/or in the presence of a continuous stress"* [4], offers a very generic and programmatic statement in order to cope in a world of greater uncertainties and systemic threats to these critical infrastructures [5].

It is now common to read about the need and call for resilience in a very wide range of publications, in areas such as globalisation processes, ecology, business strategy, urban dynamics, financial markets and personal life. Globalisation should be resilient, economies should be resilient, companies should be resilient, banks should be resilient, societies should be resilient, cities should be resilient, individuals should be resilient, etc. In safety research, the topic of resilience has also gained momentum in the past decade in particular through the thrust of authors in cognitive and system safety engineering [4].

The central idea of resilience derives from first, a deconstruction of the notion of human error, second, a better appreciation of the expertise of front-line operators (their abilities to cope with complexity) and third, a systemic view of safety, aggregating individual trade-offs to infer behaviour of complex systems. Resilience engineering is therefore an approach and practice which looks positively into people's expertise when facing daily trade-offs, and which aims to combine the aggregated effects of these behaviours at a system level to anticipate their consequences. It shares, with other research traditions, such as high reliability organisations, this methodological perspective which consists in studying the daily operations of high-risk systems and critical infrastructures (rather than focusing solely on disasters [6]). And, both these traditions face the problem of studying safety from a multi-level research.

2.2 Challenges of Multilevel Research

Why undertake multilevel research? Thirty years of research argues for the importance of this methodological problem, as for instance when conceptualising evolution of research topics from technical, human then organisational lenses [7]. If the topic is the prevention of major disruption in safety-critical systems, such as disaster caused by nuclear power plant, aircraft crash or toxic chemicals release, research strategies have to be based on interdisciplinary and multilevel principles.

Indeed, we know from reports (or experience of investigating major accidents [8]) that they are the products of strategic choices and leadership practices by top executives, organisational processes, structures and management, teams and operational actors, regulatory and inspectorate dynamics, material and sociocognitive properties as well as engineering/technological aspects [9, 10].

From these reports, we know that accidents can't be reduced, for instance, to front-line operators' activities. Focusing on their work in daily operations fails to provide the broad picture needed for understanding the construction of safety at the scale of what is revealed in retrospect in major events. Engineering design and maintenance of installations are as important, as well as the strategy of companies and its implications for daily operations but also the inspection practices of control authorities.

But each of these topics or areas of investigation (e.g., regulatory practices, company strategy, control rooms operations, technological design) is studied separately through different field of expertise, and it remains, to this date, that an in-depth investigation best illustrates, in hindsight, a multilevel research strategy and the need to consider industrial safety from a broad angle [11]. In fact, what we know of disasters with the help of these exceptional investigations is not, or rarely, matched by studies of daily operations from this multilevel perspective. There are at least three reasons for this, which turn into specific challenges.

Firstly, the resources spent to find out what happened in the aftermath of a disaster allow the collection of a vast amount of data which is not very often available otherwise in daily contexts. In these exceptional circumstances, states are often empowered to proceed with in-depth investigations, to access a diversity of findings and actors, including top actors of multinationals and agencies. Secondly, through the hiring of many consultants, practitioners and academics, such investigations can rely on expertise in a range of scientific disciplines (e.g. engineering, social sciences) in order to make sense, from a diversity of angles, of the engineering and social dimensions of the event. A true interdisciplinary strategy is often applied in this context [12].

But there is a third reason. It is much easier in retrospect to link a diversity of decisions and practices of operators, engineers, managers and regulators and to consider how they occur in relation to organisational structures, cultures and power issues which combine into specific circumstances of technological and artefacts for the accident to happen the way it happened. This problem has been defined as the *hindsight bias* by psychologists or *retrospective fallacy* by sociologists or historians.

So a multilevel safety research strategy faces at least three difficulties:

1. Time, resources and broad access to data and actors;
2. Ability to use and associate a wide range of scientific expertise;
3. A clear link between technological potential failures and multiple decisions, including top management and regulators.

These difficulties can be turned into several challenges: cognitive, institutional, methodological, theoretical and empirical.

Cognitive and institutional challenges. Rasmussen, a researcher who shaped the background for the development of resilience engineering, the conceptualisation of a multilevel safety research was precisely the intention of this researcher during the 1990s, something captured graphically by his famous sociotechnical view along with the idea of an envelope of safety [13]. But, although he was at the origin of the intellectual agenda of resilience engineering which developed subsequently, this most demanding research strategy has only been little pursued theoretically or empirically, in relation to difficulty (2) above. Rasmussen anticipated it. *"Complex,* *cross-disciplinary issues, by nature, require an extended time horizon. It takes considerable time to be familiar with the paradigms of other disciplines and often time consuming field studies are required"* [13]. I have described this as Rasmussen's *strong program for a hard problem* [14].

Studying safety across levels of a sociotechnological system was at the heart of this program and requires interdisciplinarity [14]. It is a cognitive challenge as indicated in the quote first because researchers tempted by multilevel research must become familiar with a range of domains and research traditions. It is an institutional challenge because universities favour disciplinary perspectives. *"Such studies, quite naturally, are less tempting for young professors who have to present a significant volume of publications within few years to ensure tenure"* [13].

Methodological, theoretical and empirical challenges. But there are other methodological, theoretical and empirical challenges associated that Rasmussen did not discuss in his time, especially difficulties (1) and (3). A first challenge when studying daily operations is what, who, how, when and for how long to observe, meet and connect a very wide range of category of natural events, artefacts and actors creating these complex and highly dynamic networks [10]. There is a diversity of them which contribute to the safe performance of a sociotechnological system.

Based on my experience of the chemical industry, this can potentially concern, depending on the location and size of the plant and organisation, a rather high number of differentiated natural phenomena, objects and individuals ranging from heat, cold, wind, fog, valves, pipes, chemical products to procedures, screens but also software coupled with the activities of front-line operators, site managers, corporate actors and control authorities as well as subcontracting companies including consulting ones (e.g., engineering, management). And it is the nature and quality of their interactions which produces a certain level of safety.

It is precisely through these interactions that safety is constructed on a daily basis, what is crucial is therefore to understand the results of their interactions, which represent therefore a tremendous number of them. For instance, when concentrating on human actors, all of them have a contribution at different levels, whether when participating as a safety engineer to the design phase of a project (e.g., anticipating hazardous scenario), when deactivating as an operator an alarm in a control room or when managing a team as the head of a safety department. The vast amount of interactions between these actors in their natural/atmospheric, software and material environment represents the basis of the daily construction of safety. The aggregated results of these interactions are daunting to anyone interested in the study of safety from a multilevel perspective.

It is clear that only prolonged periods of time can allow an external observer to get to appreciate these complexities, but also to make sense of them... Rain, cold weather, heat, storms, valves, chemicals, texts, diagrams, screens, formulas, tools, procedures, logs, reports, symbols, practices, operators, safety or maintenance engineers, production or site managers, actors of unions, CEO, control authority inspectors, etc. It is this mix of material and social networks which ensures reliability, resilience or safety.

This creates methodological challenges which are financial (funds available for prolonged fieldwork) and access to data combined with legal issues (especially when it comes to top decision makers). Another one is whether such an approach should be implemented by one or several researchers, one problem being to coordinate points of views when multiplying expertise and researchers.

When one can pay attention to this diversity of interconnected and distributed artefacts, objects, individuals and contexts whether climatic or institutional, one is baffled by the ability of these "ecosociotechnical" networks to remain within the boundary of safety performance given the infinite number of adaptations produced in real time and the associated flows of decisions taken. This is precisely why the solution from a research point of view is often to focus on one aspect of the problem, say, the study of process operators' interactions in a control room, the study of a maintenance team and service interactions, the study of leadership of the management of a department or the study of a chemical reaction. And, it is particularly convenient because it is precisely how some disciplines have established themselves, namely by specialising in certain themes corresponding to a degree of description of phenomena. The problem becomes one of relationship between parts and whole when involved in a multilevel study.

How to grasp the whole when one has only tools and concept for looking into parts? If one can look into the interactions of a team at the shop floor level, the conditions under which they produce their expert tradeoffs on a daily basis are products of organisational features, engineering design and strategic orientations of companies. There is therefore a need to look into the interactions of a lot of actors in the way they combine to shape these interactions. Because it is impossible to look at everything, choices must be made of who, when, where and how to look and to probe, then access to observations and interviews must be granted (which is much easier when it comes to process operators than to executives!), so the methodological challenge is also tightly connected to a theoretical one.

When one is potentially granted access to a very wide range of situations as described above, how to organise the material collected to interpret findings in relation to the topic of safety? The problem is that the amount of data produced is potentially huge and an intellectual background for organising these data is needed, which leads to the question of the availability of a model (or several) to do this. Because there are many different angles of observations related to different domains of knowledge, the model supporting the links between them is one crucial aspect which is a theoretical challenge. Graphical models have been key to help frame this complex issue [6, 9], but they can only help structure our mindset, and analytical developments are also needed. Moreover, sometimes, areas of great interest for safety are also not investigated. One example is the field of strategy decision making, for which one needs to produce specific research to better approach this important dimension from an empirical point of view.

Finally, another challenge is to deal with trends of our contemporary world of globalised activities with intense and uncertain market competitions (Sect. 2.1 above). The current operating conditions as introduced in the first section create difficulties for any research, first because organisations are not stabilised and are likely to

Cognitive challenge: knowing then combining independently developed conceptual schemes from different research traditions.

Institutional challenge: overcoming disciplinary boundaries established through institutions (universities, journals).

Methodological challenge: observing and accessing a very wide diversity of artefacts and actors shaping the conditions of safe operations within a sufficient amount of time.

Theoretical challenge: selecting, ordering and interpreting a vast amount of data in a way that is consistent with the object of study, safety, while combining different insights.

Empirical challenge: facing globalised contexts of extended chains of commodities, speed of multifaceted changes related to competitive markets and trends affecting reliability, resilience or safety of operations.

Fig. 2.1 Short description of challenges of multilevel safety research

evolve quickly (which can restrict the validity of interpretations to a limited period of time), second, because many organisations are now part of worldwide networks and information infrastructures which connect actors from multiple locations over the world, and in multiple entities. This last point also indicates the problem of understanding local realities with a view of the macro trends affecting the constraints of safety critical systems, such as standardisation, outsourcing or financialisation, to select a few of the trends of the past decades associated with globalised processes [15]. This in an area where research is needed too.

Multiple challenges of multilevel safety research. The table summarises the challenges of a multilevel safety research as briefly (more can be found in [16]) addressed in this chapter and which derives from three difficulties: time and data access (1), interdisciplinarity (2) and link between potential failures and decisions (3). These difficulties are decomposed in cognitive, institutional, methodological, empirical and theoretical challenges summarized in Fig. 2.1.

2.3   Conclusion

Resilience is a notion resonating with the current moment of history where surprises of different kinds have become part of our expectations, requiring the ability to both anticipate and react in a timely manner for a diversity of situations, including, among other, extreme natural events, terrorism, technology breakdown or financial disturbances. This contemporary situation requires from safety-critical systems an ability to adapt to uncertain and potentially fast changing environments. Resilience engineering addresses this topic, as other research traditions, such as high reliability organisations, both facing the problem of multilevel research of the complex, networked, globalised and constructed nature of safety. When following such a multilevel strategy, researchers meet cognitive, institutional, methodological, theoretical and empirical challenges.

References

1. I. Goldin, M. Mariathasan, *The Butterfly Defect: How Globalization Creates Systemic Risks, and What to Do About It* (Princeton University Press, Princeton, 2015)
2. J.-C. Le Coze, An essay: societal safety and the global[1,2,3]. Saf. Sci. (December 2018, Volume 110, Part C), https://doi.org/10.1016/j.ssci.2017.09.008
3. L.K. Comfort, A. Boin, C.C. Demchak, Resilience revisited - an action agenda for managing extreme events, in *Designing Resilience: Preparing for Extreme Events* (University of Pittsburgh Press, Pittsburgh, 2010)
4. E. Hollnagel, D.D. Woods, N. Leveson (eds.), *Resilience Engineering: Concepts and Precepts* (Ashgate, Aldershot, 2006)
5. C. Perrow, *The Next Catastrophe: Reducing our Vulnerabilities to Natural, Industrial and Terrorist Disasters* (Princeton University Press, Princeton, 2011)
6. J.-C. Le Coze, Vive la diversité! High reliability organisation (HRO) and resilience engineering (RE). Saf. Sci. (2016, In press)
7. A.R. Hale, J. Hovden, Management and culture: the third age of safety. A review of approaches to organizational aspects of safety, health and environment, in *Occupational Injury. Risk Prevention and Intervention*, ed. by A.M. Feyer, A. Williamson (Taylor & Francis, London, 1998)
8. J.-C. Le Coze, Accident in a French dynamite factory: an example of an organisational investigation. Saf. Sci. **48**(1), 80–90 (2010)
9. J.-C. Le Coze, New models for new times. An anti dualist move. Saf. Sci. **59**, 200–218 (2013)
10. J.-C. Le Coze, Outlines of a sensitising model for industrial safety assessment. Saf. Sci. **51**(1), 187–201 (2013)
11. A. Hopkins, *Disastrous Decisions: The Human and Organisational Causes of the Gulf of Mexico Blowout* (CCH Australia, 2012)
12. J.-C. Le Coze, Disasters and organisations: from lessons learnt to theorising. Saf. Sci. **46**(1), 132–149 (2008)
13. J. Rasmussen, Risk management in a dynamic society: a modelling problem. Saf. Sci. **27**(2), 183–213 (1997)
14. J.-C. Le Coze, Reflecting on Jens Rasmussen's legacy. A strong program for a hard problem. Saf. Sci. **71**, 123–141 (2015)
15. J.-C. Le Coze, Globalization and high-risk systems. Policy Pract. Health Saf. **15**(1), 57–81 (2017)
16. J.-C. Le Coze, *Trente ans d'accidents: Le nouveau visage des risques sociotechnologiques* (Octarès, 2016)

Chapter 3
Moments of Resilience: Time, Space and the Organisation of Safety in Complex Sociotechnical Systems

Carl Macrae

Abstract When and where does resilience happen? This is one of the most funda-mental issues in the theory and practice of resilience in complex systems. Is resilience primarily a reactive or a proactive organisational property? Does it emerge locally and rapidly, or over longer time periods and at larger scales? This chapter develops a framework that seeks to characterise how resilience unfolds at three different scales of organisational activity: situated, structural and systemic. This analysis highlights the importance of understanding how locally situated activities of adjustment and recovery can trigger more generalised structural reforms, and how these might sup-port wide-ranging systemic reconfigurations across entire industries. This analysis draws on practical examples from aviation, healthcare and finance.

Keywords Disruption · Adaptation · Reorganisation · Learning · Risk

3.1 Understanding Resilience: When and Where?

Many current efforts to understand and manage risk in complex organisations focus on ideas—and ideals—of resilience. Resilience has been theorised in a variety of different and sometimes conflicting ways, but broadly refers to the capacity of a system to handle disruptions, failures and surprises in ways that avoid total system collapse—and may lead to adaptation and improvement. This chapter addresses one of the most enduring challenges faced in both the theory and practice of resilience in complex sociotechnical systems: 'when' and 'where' does resilience occur? Do activities of resilience occur solely in response to adverse events, or are they pre-emptive and proactive? Is resilience characterised by rapid processes of adjustment

---

C. Macrae (✉)
Centre for Health Innovation, Leadership and Learning, Nottingham University Business School, University of Nottingham, Nottingham, UK
e-mail: carlmacrae@mac.com

© The Author(s) 2019
S. Wiig and B. Fahlbruch (eds.), *Exploring Resilience*, SpringerBriefs in Safety Management, https://doi.org/10.1007/978-3-030-03189-3_3

that unfold over minutes and hours, or long-term reorganisations that take years and decades? And does resilience primarily emerge through the activities of those working on the operational frontline, or through higher-order processes that span entire industries? These questions of time and space are fundamental to how we understand and operationalise resilience. However, to date, these issues have largely been assumed rather than explored. Rather like the broader literature on risk [1], the current literature on resilience represents something of an archipelago. Many small islands of research each examine resilience at different scales of activity, with few systematic attempts to examine the linkages between these. This chapter examines these issues and presents a framework for understanding resilience across different scales of organisational activity and considers the key implications for theory and practice.

Our current theories of resilience address issues of time and space in different ways. Regarding the temporal question of 'when' resilience happens, theories differ as to whether resilience happens before or after a disruptive event, and either quickly or slowly [2]. Some emphasise that resilience is solely a reactive capacity, characterised by efforts to respond, recover and repair once disruptive events have occurred [3]. Other characterisations expand the temporal reach of resilience, to "the intrinsic ability of a system to adjust its functioning before, during, or after changes and disturbances, so that it can sustain required operations under both expected and unexpected conditions" [4]. Likewise, Comfort, Boin and Demchak [2] define resilience as "the capacity of a social system [...] to proactively adapt to and recover from disturbances that are perceived within the system to fall outside the range of normal and expected disturbances". More recent analyses have emphasised the need to distinguish between distinct forms of 'precursor' resilience, that proactively prevents major system failures occurring, and 'recovery' resilience, that rapidly responds after a system collapse [5]. Relatedly, regarding the spatial question of 'where' resilience happens, theories diverge on the location and source of resilience in complex sociotechnical systems. The predominant focus is on the adaptive capacity of frontline personnel who encounter and manage immediate fluctuations in organisational activity [6, 7]. Other analyses focus on specialist supervisory professionals that both oversee and retain close contact with the frontline [5, 8, 9]. Alternative approaches emphasise the role of supra-organisational regulatory bodies [10, 11] or the interconnected capacities of entire social and economic systems [3, 12].

3.2 Moments of Resilience: Situated, Structural and Systemic

To understand resilience at different scales of time and space, this chapter introduces a framework that characterises resilience in terms of the *scale and nature of organisational activity that unfolds around a disruption*. This framework characterises organisational activities as unfolding within three broad "moments" of resilience:

situated, structural and systemic. Each of these moments represents a different scale of organisational activity in terms of duration and reach across a system. Each also represents a different way of enrolling core sociotechnical resources—such as knowledge, tools, data, skills and ideas—into organisational activities.

- *Situated resilience* emerges at or close to the operational frontline. It involves mobilising and combining existing sociotechnical resources to detect, adjust to and recover from disruptive events. This can unfold over seconds to weeks.
- *Structural resilience* emerges in the monitoring of operational activities. It involves the purposeful redesign and restructuring of sociotechnical resources to adapt to or accommodate disruptive events. This can unfold over weeks to years.
- *Systemic resilience* emerges in the oversight of system structure and interaction. It involves reconfiguring or entirely reformulating how sociotechnical resources are designed, produced and circulated. This can unfold over months to decades.

To develop and illustrate this framework, this chapter draws on comparative analysis of three diverse sectors—healthcare, aviation and finance. These sectors differ considerably in terms of institutional landscape, operational practice and social organisation, as well as the nature of the risks to be managed. This helps illustrate both differences and similarities in how resilience is operationalised.

3.3   Situated Resilience

Situated resilience emerges from the situated practices that unfold around disruptive events, close to the operational frontline, and involves integrating and applying existing sociotechnical resources such as knowledge, data, tools and skills to detect and respond to disruptive events as they occur. Moments of situated resilience represent organisational activity at a micro-level: the dynamic interactions of people and their immediate work environment, and the adaptation, adjustment and intelligence required to handle unexpected and non-routine events in front-line work [13, 14] by mobilising the requisite sociotechnical resources. Situated resilience can involve the rapid detection and resolution of deviations from plans. For instance, in airline operations, incorrect departure route data in the flight computer may be missed during routine cross-checks, leading to an unintended departure route being flown. Ongoing monitoring by air traffic controllers allows unexpected route deviations to be detected and rapidly addressed by calling the aircraft and providing a corrected route. A sequence such as this, lasting barely a few minutes, represents a disruption to intended activity that requires multiple actors deploying existing resources to recover intended operations.

Situated resilience can involve rapidly responding to and organising around rare emergency events. In maternity care in healthcare, for instance, emergencies—such as post-partum haemorrhage—are relatively rare but require immediate lifesaving action. This involves rapidly mobilising and applying specialist knowledge, skills and tools in patterns of interaction to diagnose and treat the emergency, the precise features of which have probably never been encountered by this team in this specific combination. And situated resilience can involve the instituting of practices that create spaces that support the detection and recovery from hidden problems. In financial services, for instance, it is customary for front office staff (those with trading or related responsibilities) to take at least a two week continuous break from work each year, handing over their trading book to a colleague, in part to provide an opportunity for any irregularities to be identified. Similarly, deal teams involved in large and complex transactions, such as multi-billion dollar purchases of infrastructure assets, may spend weeks preparing alternative, back-up deal documents to help accommodate last-minute changes in deal terms that could threaten the transaction.

3.4 Structural Resilience

Structural resilience represents the processes of restructuring and reforming sociotechnical resources and situated practices. These processes can span multiple organisational units and are typically coordinated by groups that monitor and supervise frontline operational activities. Moments of structural resilience represent organisational activity at the meso-level: active processes of examining organisational practices and sociotechnical resources, and redesigning them in light of past experiences [15], through an effortful structuration process seeking to shape both situated practice and social structure [16]. Structural resilience can involve the reorganisation of work systems in response to disruptive events. For example, an airline's event involving a high-speed rejected take-off due to a spurious engine fire warning was revealed, after detailed investigation, to be due to loose screws on a temperature sensor. During maintenance, all screws had been loosened by a new engineer following company procedure. But after a handover, a different engineer finishing the job only tightened the screws that were part of the local informal approach to the task. Revealing this gap between plans and practices [15] or work as done and work as imagined [17] allowed the work practices, the cultural norms and the formal processes to be restructured and reformed over several months.

Structural resilience can also involve the design and redesign of sociotechnical resources through the simulation of disruptive events. In healthcare, the on-site or 'in situ' simulation of obstetric emergencies is used not only to train individual and team skills, but can also support the redesign and restructuring of the wider sociotechnical infrastructure and resources that support effective responses to emergency events. For instance, regular in situ simulation of emergencies such as major haemorrhage allows the continual testing and improvement of the design of decision aids, the accessibility of equipment and the processes for requesting and receiving blood products. Similar mechanisms of structural resilience are organised across multiple organisations in financial services. For example, regular 'stress tests' of financial institutions simulate extreme adverse scenarios, and are used to assess the stability and safety of current resources and structures, and adapt them where necessary [18]. Structural resilience can also involve the restructuring of resources when indicators of potential risk are triggered. In finance, countercyclical capital buffers require firms to build up additional capital reserves during periods of credit expansion ("good times"), both to better prepare for unexpected losses in times of financial stress and modulate risk taking.

3.5   Systemic Resilience

Systemic resilience represents the fundamental reconfiguration and reform of the processes that design, produce, constitute and circulate the sociotechnical resources that underpin safety. This can take place over years to decades, enrol large numbers of actors and cross many boundaries across an entire industry. Moments of systemic resilience represent organisational activity at the macro-level: lengthy, elaborate and often heavily contested negotiations regarding the proper configuration of sociotechnical resources, the appropriate means of generating these and the systems that supports these. Systemic resilience can involve wide-scale reform of the assumptions, norms and technological systems underlying activities across a sector, reconfiguring the way that disruption is itself handled. For example, in aviation the disturbing failure to trace and recover wreckage or critical flight data from MH370, a Boeing 777 lost in 2014, has provoked a fundamental reconfiguration of the way aircraft data are traced and recovered globally. This process is unfolding over years and represents the global aviation system slowly reconfiguring and adapting in response to a serious systemic disruption.

Systemic resilience can also involve considerable reconfigurations of the system-wide architecture for detecting and responding to disruptions. In healthcare, for example, a major crisis centred on sustained failures of care at a UK hospital in Mid Staffordshire prompted, through a multi-year public inquiry, a dramatic reshaping of the function of system-wide regulatory and inspection processes [19]. This fundamentally reconfigured the system-wide mechanisms for detecting and uncovering similar problems. A related reconfiguration has involved the creation of an independent and system-wide investigation body, to conduct blame-free and systems-focused investigations [20, 21]. Similar reconfigurations took place in financial services following the financial crisis of 2008–2009. This included the design and introduction of system-wide countercyclical capital buffers, previously discussed, representing a new means of detecting, measuring and managing one of the sources of the prior crisis. These all represent fundamental reformulations of the core assumptions and systemic architectures that produce and shape sociotechnical resources, in response to serious systemic crises.

3.6 Organising Resilience: From Disruption
to Reconfiguration

Resilience can be understood as happening both quickly and slowly, as a multi-layered set of processes enacted over different time periods and over different scales of activity. Distinguishing three broad 'moments' of resilience, each with a distinct function and logic, raises key questions with practical and theoretical implications. First, when does resilience occur? Previously, this question has been answered with reference to some materially disruptive event—either proactive action 'before', or reactive action 'after'. However, disruptions that provoke resilience can simply be symbolically, rather than materially, disruptive [9, 22]. Resilience can be provoked by disruptions to expectations, assumptions, norms and beliefs that call into question the safety of current organisational activity. This removes the need to define resilience directly in relation to a materially adverse outcome. Simulating 'imagined' emergency scenarios to test and adapt systems provides a clear and direct example of this [23]. However, reliably generating and responding to symbolic disruptions can be challenging. In some sectors, materially adverse events—such as air accidents—necessarily provoke dramatic symbolic disruption: planes are not supposed to crash. But in other sectors, like healthcare and finance, materially adverse outcomes are an expected part of organisational life. Patients are ill and sometimes do not survive. Creditors go bad or the market can turn. In many circumstances, death and losses need not provoke surprise or symbolic disruption and may be normalised. This emphasises the importance of the difficult, effortful interpretive work that must be done to actively construct and communicate the symbolic disruptions that can act as provocations for resilience across different scales of organisational activity. Building resilient systems therefore depends, in part, on building an infrastructure that can not only detect and respond to materially adverse events, but can continually manufacture, enlarge and circulate symbolically disruptive events to organise resilience around: surprises, uncertainties, ambiguities and other challenges to current norms and beliefs. This suggests that, to support resilience, industries need mechanisms—and people—at every level of the system that can generate scale-appropriate symbolic disruptions that provoke resilient, adaptive responses.

Second, organisational life is full of fluctuation, variation and interruption. But when does a mere fluctuation become a disruption, and how does this lead to the enactment of situated, structural or systemic resilience? A defining feature in this analysis is that a disruption is a 'disruptive interruption': it interrupts an activity in such a way that it derails the ongoing flow of that activity and requires the mobilisation of supplementary sociotechnical resources (e.g. expertise, attention, time, tools, data) to restore order and control, beyond those that would ordinarily be enrolled in that particular activity. When defined this way, disruption is scale-insensitive. It is equally relevant to the situated practice of frontline workers as it is to the systemic reorganisation of entire industries. If this is the case, how do disruptions at one scale of activity migrate and enlarge to enact resilience at greater scales of activity? This analysis suggests that the enactment of resilience across different 'moments' is, in part, dependent on 'scaling-up' a perceived disruption. For fluctuations to become disruptions and provoke situated resilience, they need to represent a *perceived loss of situated control*: a failure of current situated practice to maintain control and comprehension of activities that creates a perceived need to activate additional sociotechnical resources to re-establish control. Likewise, to enact structural resilience, disruptions need to represent a *perceived structural collapse*: a failure in the performance, design or functioning of current structures and resources, that requires purposeful restructuring. And to enact systemic resilience, disruptions need to pose a *perceived systemic crisis*: a failure of current system-wide arrangements to properly supply the resources needed for effective control and functioning of the system, requiring a broad-based reconfiguration.

In practice, this suggests that operationalising resilience across different moments and scales of activity requires protected spaces and forums that create vertical alignment within industries: spaces in which more local disruptions can be transformed into more expanded, larger-scale disruptions. There is a risk that expanding a disruption can imply expanding blame. Thus, these spaces need to be protected from contests over liability and be removed from pressures to allocate or deflect blame. Likewise, expanding disruption across time and space requires expanded scales of expertise. Industries need cadres of professionals that work at the interfaces of situated activities, structural supervision and systemic oversight, who are adept at making linkages between these different levels of analysis—and at constructing and communicating compelling symbolic disruptions. Organisational safety units [9, 24], independent accident investigators [25] and reliability professionals [5] provide potential models for these protected spaces and professional groups.

It can be hard to see the relationship between momentary seconds of operational activity, and years-long reconfigurations across entire industries. There remains much work to be done to explain and operationalise resilience at different moments and scales of organisational activity. Future efforts might most productively focus on three areas: the transition from normal fluctuation to provocative disruption; the interfaces between different scales of resilient activity; and the nature of resilience as it unfolds in situated, structural and systemic ways. Resilience can be both fast and slow, small and large. Building more resilient systems depends on being able to conceptually pull apart and practically integrate these moments of resilience.

References

1. C. Hood, D.K.C. Jones (eds.), *Accident and Design: Contemporary Debates on Risk Management* (Routledge, London, 1996)
2. L.K. Comfort, A. Boin, C.C. Demchak (eds.), *Designing Resilience: Preparing for Extreme Events* (University of Pittsburgh Press, Pittsburgh, 2010)
3. A. Wildavsky, *Searching for Safety* (Transaction Books, New Brunswick, 1988)
4. E. Hollnagel, Proactive approaches to safety management. The Health Foundation thought paper (2012)

5. E. Roe, P.R. Schulman, *Reliability and Risk: The Challenge of Managing Interconnected Infrastructures* (Stanford University Press, Stanford, 2016)
6. E. Hollnagel, R.L. Wears, J. Braithwaite, From Safety-I to Safety-II: A white paper. Technical report, The Resilient Health Care Net (2015)
7. K.E. Weick, K.M. Sutcliffe, D. Obstfeld, Organizing for high reliability: processes of collective mindfulness. Res. Organ. Behav. **21**, 81–123 (1999)
8. C. Macrae, Learning from patient safety incidents: creating participative risk regulation in healthcare. Health Risk Soc. **10**(1), 53–67 (2008)
9. C. Macrae, *Close Calls: Managing Risk and Resilience in Airline Flight Safety* (Palgrave Macmillan, New York, 2014)
10. C. Macrae, Regulating resilience? Regulatory work in high-risk arenas, in *Anticipating Risks and Organising Risk Regulation*, ed. by B.M. Hutter (Cambridge University Press, Cambridge, 2010)
11. C. Macrae, Reconciling regulation and resilience in health care, in *Resilient Health Care*, ed. by E. Hollnagel, J. Braithwaite, R.L. Wears (Ashgate, Farnham, 2013)
12. J. Rodin, *The Resilience Dividend: Being Strong in a World Where Things Can Go Wrong* (Public Affairs, New York, 2014)
13. E. Hutchins, *Cognition in the Wild* (MIT Press, Cambridge, 1995)
14. L.A. Suchman, *Plans and Situated Actions: The Problem of Human-Machine Communication* (Cambridge University Press, Cambridge, 1987)
15. R. Miettinen, J. Virkkunen, Epistemic objects, artefacts and organizational change. Organization **12**(3), 437–456 (2005)
16. A. Giddens, *The Constitution of Society: Outline of the Theory of Structuration* (University of California Press, Berkeley, 1984)
17. E. Hollnagel, D.D. Woods, Epilogue: resilience engineering precepts, in *Resilience Engineering: Concepts and Precepts*, ed. by E. Hollnagel, D.D. Woods, N. Leveson (Ashgate, Aldershot, 2006)
18. Bank of England, Stress testing the UK banking system: key elements of the 2016 stress test. Technical report, Bank of England (2017)
19. R. Francis, *Report of the Mid Staffordshire NHS Foundation Trust Public Inquiry* (The Stationery Office, 2013)
20. C. Macrae, C. Vincent, Learning from failure: the need for independent safety investigation in healthcare. J. R. Soc. Med. **107**(11), 439–443 (2014)
21. PASC, Investigating clinical incidents in the NHS. UK House of Commons Public Administration Select Committee report (2015)
22. B.A. Turner, The organizational and interorganizational development of disasters. Adm. Sci. Q. **21**(3), 378–397 (1976)
23. C. Macrae, T. Draycott, Delivering high reliability in maternity care: in situ simulation as a source of organisational resilience. Saf. Sci. (2016, In press)
24. P.R. Schulman, The negotiated order of organizational reliability. Adm. Soc. **25**(3), 353–372 (1993)
25. C. Macrae, C. Vincent, Investigating for improvement: building a national safety investigator for healthcare (2017). CHFG thought paper

Chapter 4
Resilience Engineering as a Quality Improvement Method in Healthcare

Janet E. Anderson, A. J. Ross, J. Back, M. Duncan and P. Jaye

Abstract Current approaches to quality improvement rely on the identification of past problems through incident reporting and audits or the use of Lean principles to eliminate waste, to identify how to improve quality. In contrast, Resilience Engineering (RE) is based on insights from complexity science, and quality results from clinicians' ability to adapt safely to difficult situations, such as a surge in patient numbers, missing equipment or difficult unforeseen physiological problems. Progress in applying these insights to improve quality has been slow, despite the theoretical developments. In this chapter we describe a study in the Emergency Department of a large hospital in which we used RE principles to identify opportunities for quality improvement interventions. In depth observational fieldwork and interviews with clinicians were used to gather data about the key challenges faced, the misalignments between demand and capacity, adaptations that were required, and the four resilience abilities: responding, monitoring, anticipating and learning. Data were transcribed and used to write extended resilience narratives describing the work system. The narratives were analysed thematically using a combined deductive/inductive approach. A structured process was then used to identify potential interventions to improve quality. We describe one intervention to improve monitoring of patient flow and organisational learning about patient flow interventions. The approach we describe is challenging and requires close collaboration with clinicians to ensure accurate results. We found that using RE principles to improve quality is feasible and results in a focus on strengthening processes and supporting the challenges that clinicians face in their daily work.

J. E. Anderson (✉) · J. Back · M. Duncan
Florence Nightingale Faculty of Nursing, Midwifery and Palliative Care,
King's College London, London, UK
e-mail: janet.anderson@kcl.ac.uk

A. J. Ross
Dental School, School of Medicine, University of Glasgow, Glasgow, Scotland, UK

P. Jaye
Guy's and St. Thomas' NHS Foundation Trust, London, UK

© The Author(s) 2019
S. Wiig and B. Fahlbruch (eds.), *Exploring Resilience*, SpringerBriefs
in Safety Management, https://doi.org/10.1007/978-3-030-03189-3_4

Keywords Emergency department · Patient flow · Adaptive capacity · Quality improvement

4.1 Context and Introduction

Resilience Engineering (RE) is a new paradigm for conceptualising how work is accomplished in complex adaptive systems such as healthcare [1, 2]. It explicitly argues that the ability of organisations to adapt to pressures is what makes the system work, and is responsible for maintaining good outcomes in spite of problems and challenges. Workers are therefore seen as the key to creating safety, rather than being cast as the weak link in the system, prone to error and responsible for adverse outcomes. RE argues that it is the variability in the healthcare environment that drives the need for adaptation [3]. For example, surges in patient numbers, multiple patients deteriorating at the same time, lack of equipment and inappropriate staffing are all common variations in the conditions of work that require adaptation by workers. This way of thinking is different to the assumptions underpinning most quality improvement efforts that attempt to constrain human behaviour by specifying via protocol what actions should be taken [4], based on past problems identified through incident reporting, audits, or identification of waste through Lean principles.

These ideas appeal to clinicians and safety researchers because they reflect the reality of the messy clinical world in which conditions cannot always be anticipated and solutions have to be improvised. However, they need further interpretation and elaboration to move from a description of how work is achieved, to inform quality improvement [5]. RE is a theory about systems, and it needs to move beyond individual adaptations to consider how a system might support adaptive capacity. The four resilience abilities of responding, monitoring, anticipating and learning, proposed by Hollnagel [6], are promising and could provide a means for thinking about how adaptive capacity can be supported. For example, by considering whether and how a system learns it might be possible to devise ways to enhance learning and thereby increase quality. Despite these promising concepts, it is not immediately clear how to define the focus of an investigation based on RE since simply targeting learning in general, for example, seems unlikely to have a measurable effect on outcomes of interest.

In this research we developed a conceptual model to help us to think about how quality can be improved using insights from RE [4]. The CARe model proposes that variability in the healthcare environment often occurs because of a mismatch between demand and capacity. For example, a surge in patients is a problem if there are not enough staff rostered. Demand-capacity misalignments lead to adaptations in situ as staff attempt to work around problems to deliver care. Outcomes emerge from the interplay of misalignments and adaptations. A key insight from the model is that there are two potential routes to improving quality. Improvement efforts could focus on reducing misalignments between demand and capacity, thereby reducing the need for adaptations. This could potentially preserve resources that would otherwise be used to solve problems that have an obvious standardised solution (such as ensuring there is a good system for maintaining equipment) so that they could be used for coping with other less predictable problems. Alternatively, better support for adaptations and for strengthening the link between adaptations and good outcomes could also be a way to improve quality. Adaptations carry the risk that they will result in adverse outcomes because people are departing from protocol, or improvising solutions to problems not covered by the protocol, and may not be able to foresee all the implications of their actions. Supporting adaptation to ensure a good outcome is one goal of quality improvement from an RE perspective. For example, better systems for monitoring risk might be of use in enabling better planned adaptations when there are high risk conditions.

We have used RE theory and the CARe model to investigate whether RE can be used as a quality improvement method. Working longitudinally over several years, we have studied in depth the work systems the Emergency Department in a large London teaching hospital. The overall aims of the research were to use RE theory to develop and evaluate quality improvement interventions. To do this, we aimed to:

1. Build a deep and nuanced understanding of how work was achieved in the two units, including misalignments between demand and capacity and adaptations performed in situ;
2. Develop an interpretive process to identify interventions;
3. Design interventions with clinical teams and implement them;
4. Evaluate outcomes.

4.2   Methodology

For the ethnographic field work, two researchers, working as non-participant observers, first identified the main staff roles, processes, co-ordinating mechanisms, such as meetings and handovers, and technology and tools used. More focused observations were then conducted of the co-ordinating mechanisms and these included staff and team meetings, ward rounds, board rounds, patient flow meetings, and handovers. Finally, staff were shadowed as they carried out their everyday work and were asked to clarify decision making processes and reasons for actions. In depth interviews (n = 13) were also conducted with staff to probe for further detail about phenomena observed and clarify researchers' understanding of the observed work. Observational work occurred in both units concurrently.

Fieldwork data (104 h of observation) were captured in written form and transcribed to electronic format. Field notes were expanded upon, combined with interview data, and then used as the basis for writing extended resilience narratives describing how outcomes emerge from the interplay of misalignments and adaptations. The aim was to describe trajectories of action that would serve as the basis for identifying opportunities for intervention. The resilience narratives were then analysed thematically using a combined deductive/inductive approach. Specifically, the analytic themes were – misalignments and pressures, variability, adjustments and adaptations, outcomes, goal trade-offs, anticipating, monitoring, responding and learning. The output of the analysis was a comprehensive description of the work system from the perspective of RE theory. At all stages of data collection and analysis we discussed and tested emerging findings within the research team, including clinicians, and with a clinical advisory group in each unit.

We then developed a structured collaborative process to design and implement interventions. The researchers subsequently developed a series of intervention proposals based on the ethnographic results. Clinical staff attended a series of workshops to discuss the results and advise on which interventions were most feasible and relevant. The design and implementation of the interventions was then conducted with the clinical partners who were most knowledgeable and influential in each unit.

4.3   Results

In the UK at the time the study was carried out, emergency departments were required to treat and discharge 95% of patients within four hours. Preventing breaches of this target was therefore a major focus of quality efforts. Regular patient flow meetings were held every two hours in the department, convened by a patient flow co-ordinator, to review patient numbers at all points in the department, flow through the department and to trouble shoot potential breaches of the waiting time target. Immediately before the meeting the patient flow co-ordinator would manually tally numbers of patients at various points in the department and verbally ascertain from clinicians which patients were likely to be imminently discharged and for those who were not, identify what was causing delay(s). Discussion at the meeting focused on how to address any particular problems and avert breaches, and often involved decisions to flexibly reallocate staff to different areas. Observations showed that each new meeting started with a new tally of patient numbers and did not refer to the actions recommended at the previous meeting two hours ago. Thus, it was not possible for staff to know;

- Whether the recommended action had been implemented;
- What the intended effect of the action was;
- What effect the action had in practice.

For staff reallocated to an area, it was not clear how long they were to remain and what they were trying to achieve. However, in some cases it was obvious. For example, if triage was overwhelmed with many simultaneous arrivals, a nurse flexed to this area would focus on reducing the numbers waiting. But there was no feedback to the flow co-ordinator and the next meeting would begin by reviewing numbers in each area with no reference to previous actions suggested.

In RE terms this resulted in an inability to monitor both the recommended action and its outcome, and an inability to learn from previous actions when convening the next two hourly meeting. The intervention that we developed involved redesigning the document used and the procedure for the meeting. The form was redesigned to enable capture of recommended actions and intended outcomes. The redesigned meeting process involved starting the meeting by reviewing actions from the previous meeting and evaluating whether they had had the desired effect. Decisions could then be made to address any problems that hadn't been solved in the previous meeting in a new way, before moving on to consider any additional problems that had developed in the previous two hours. These interventions aimed to increase the capacity of the patient flow meetings to monitor and learn from actions taken to improve patient flow in order to increase the adaptive capacity of the system.

4.4 Discussion

In this work we have demonstrated that RE can be used to identify opportunities to improve quality and to develop quality improvement interventions. In the rest of this section, we discuss some of the difficult issues and challenges faced in using RE to improve quality.

The intervention described here was designed to better support adaptive processes (adapting to patient inflow) and increase the likelihood that adaptations will lead to success (maintaining patient flow metrics). The method that we used focused attention on processes that could be strengthened to better support the challenges that clinicians had to resolve. Other quality improvement methods have different ways of identifying the targets of improvement efforts. For example, Lean approaches focus on identifying waste and intervening to reduce it and eliminate variation (for example [7, 8]). Traditional quality improvement work often starts with reported adverse incidents which indicate that the system has produced unsuccessful outcomes [9]. However, targeting the causes of previous adverse incidents carries the risk of devising futile interventions for problems that would never occur again, and conversely, not addressing other system weaknesses that have yet to cause an adverse incident.

The process we developed was challenging, partly because it entailed an iterative sense making process involving interpretation using theory and observational data. For non clinicians it was challenging to understand all the nuances of the observations and clinical partners in the research team were crucial for ensuring that our interpretations and emerging results were accurate. The challenges included; steep learning curve for researchers; prolonged data collection time; effective analysis of a large amount of data; ensuring clinical engagement. However, many of these challenges apply to most qualitative health services research and are not insurmountable.

We did not start this study with an already identified quality problem that we wanted to solve. Instead, we used RE theory to understand in depth how the work system operated and where it could be strengthened. Nevertheless, the general approach used here could also be used to address a known problem, and indeed it may be much easier to achieve as the focus would be well defined from the outset. For example, improving medication errors in a hospital ward may be an appropriate aim of this approach. In this case, RE could provide a useful adjunct to existing quality improvement efforts by building a thorough understanding of work as done, misalignments between demand and capacity, sources of variability and the four resilience abilities in relation to medication administration. This would provide a thorough understanding on which to base the design of investigations and interventions. Without such a deep understanding of the system it may be difficult to design interventions that will be workable, sustainable and effective.

Evaluating quality improvement interventions based on RE is likely to be difficult. One challenge with evaluation is demonstrating that interventions increase resilience. Because adaptive capacity is expressed by a system in relation to a pressure or problem, we view it as an emergent property of the system rather than an outcome that can be measured [4]. For this reason we have not attempted to measure resilience. Our interventions have instead targeted the four resilience abilities (anticipating, monitoring, responding, learning) inferring that supporting these abilities will increase adaptive capacity. However, evaluating whether interventions have changed these abilities is also challenging and requires in depth qualitative work to understand the degree to which these abilities are affected. One concern is that interventions to strengthen processes are likely to be weakly linked to clinical outcomes and therefore it may be difficult to find strong evidence of effectiveness. This is a common problem in quality improvement that aims to change organisational processes [10, 11] and it can be particularly difficult to show that adverse incidents have been prevented.

4.5   Further Development

The approach that we have developed to quality improvement is resource intensive and required a well-grounded understanding of RE theory and practice. If this approach is to be useful in healthcare there is a need to produce guidance, streamline the process and more clearly articulate how to move from data collection to interpretation to intervention and evaluation. We are confident this can be done, but there is still a need to test the approach in a variety of settings. Primary care and mental health care are two settings in which this approach may be particularly valuable as both are less structured than acute care settings and rely to a greater extent on processes of social co-ordination and articulation that are even less amenable than acute care processes to standardisation and protocols.

References

1. E. Hollnagel, J. Braithwaite, R.L. Wears (eds.), *Resilient Health Care* (Ashgate, Farnham, 2013)
2. A. Ross, J.E. Anderson, Mobilizing resilience by monitoring the right things for the right people at the right time, in *Resilient Health Care Volume 2: The Resilience of Everyday Clinical Work*, ed. by R.L. Wears, E. Hollnagel, J. Braithwaite (Ashgate, Farnham, 2015), pp. 235–248

 3. R.L. Wears, E. Hollnagel, J. Braithwaite (eds.), *Resilient Health Care Volume 2: The Resilience of Everyday Clinical Work* (Ashgate, Farnham, 2015)
 4. J.E. Anderson, A.J. Ross, J. Back, M. Duncan, P. Snell, K. Walsh, P. Jaye, Implementing resilience engineering for healthcare quality improvement using the CARE model: a feasibility study protocol. Pilot Feasibility Stud. **2**(61) (2016)
 5. J.E. Anderson, A.J. Ross, P. Jaye, Resilience engineering in healthcare: moving from epistemology to theory and practice, in *Proceedings of the Fifth Resilience Engineering Symposium* (2013). Resilience Engineering Association
 6. E. Hollnagel, The four cornerstones of resilience engineering, in *Resilience Engineering Perspectives, Volume 2: Preparation and Restoration*, ed. by C.P. Nemeth, E. Hollnagel, S. Dekker (Ashgate, Farnham, 2009), p. 117–134
 7. Z.J. Radnor, M. Holweg, J. Waring, Lean in healthcare: the unfilled promise? Soc. Sci. Med. **74**(3), 364–371 (2012)
 8. B.B. Poksinska, M. Fialkowska-Filipek, J. Engström, Does lean healthcare improve patient satisfaction? A mixed-method investigation into primary care. BMJ Qual. Saf. **26**(2) (2016)
 9. E. Hollnagel, *Safety-I and Safety-II: The Past and Future of Safety Management* (Ashgate, Farnham, 2014)
10. J. Øvretveit, D. Gustafson, Evaluation of quality improvement programmes. Qual. Saf. Health Care **11**(3), 270–275 (2002)
11. J. Benn, S. Burnett, A. Parand, A. Pinto, S. Iskander, C. Vincent, Studying large-scale programmes to improve patient safety in whole care systems: challenges for research. Soc. Sci. Med. **69**(12), 1767–1776 (2009)

Chapter 5
Resilience and Essential Public Infrastructure

Michael Baram

Abstract This chapter begins with a commentary on resilience as the meta-concept for organizational preparedness for disruptive events, and the factors that influence the implementation of a resilience agenda. This is followed by an analysis of resilience in the special context of essential public infrastructure wherein priority is given to reliability and continuity of service, and interdependencies between infrastructures must be dealt with. The resilience agenda of a major public water supply system is then presented to illustrate the broad range of initiatives needed to ensure its resilience. Finally, policy issues are discussed regarding adaptations of resilience to meet concerns about security and sustainability.

Keywords Public infrastructure · Resilience · Reliability · Redundancy
Public water supply · Interdependencies · Safety management

5.1 Commentary on Resilience

Resilience is a term commonly used to denote the quality of an organization, structure or system that enables it to resist and recover from disruptive events [1]. As an objective, resilience takes on additional meaning in accordance with the task at hand. Most often, the task is strategic preparedness of a company or other organization for foreseeable types of disruptive events, such as flood, loss of process control, or act of terrorism. Haimes has aptly defined resilience for this task from a systems engineering perspective: "resilience represents the ability of the system to withstand a disruption within acceptable degradation parameters and to recover with acceptable losses and time limits" [2].

M. Baram (✉)
Boston University Law School, Boston, MA, USA
e-mail: mbaram@bu.edu

© The Author(s) 2019
S. Wiig and B. Fahlbruch (eds.), *Exploring Resilience*, SpringerBriefs
in Safety Management, https://doi.org/10.1007/978-3-030-03189-3_5

Other concepts lack the strategic breadth or coherence of a resilience-centered approach. The reliability concept, for example, emphasizes redundancy for bypassing potential points of failure in order to maintain continuity of operations. And a risk-based approach dedicated to quantification of risks provides a module of probabilistic information that must be subsequently grounded in a pragmatic organizational strategy. Indeed, conflicts may arise when implementing such concepts separately, such as when improving reliability of operations involves fuel storage on site and thereby creates new risks.

Thus, an organization that seeks to develop a coherent approach to disruptive events can adopt resilience as its meta-concept because it encompasses the many other "R word" concepts for addressing disruptive events: resistance, robustness, reliability, redundancy, risk analysis, risk management, recovery, and restoration [3].

The process of implementing a resilience strategy, and the practices and outcomes, will be shaped by a combination of circumstances, dynamic conditions, and lessons learned from experience that is unique for each organization. For example, an organization that has experienced disruptive events and knows its vulnerabilities, may bypass risk analysis and pragmatically focus on improving the robustness of its infrastructure and striving to prevent those events known to have the types of impacts that would destroy critical parts of its infrastructure [4].

A major factor shaping an organization's approach to resilience is its safety management system. Disruptions caused by external sources such as a Tsunami or act of terrorism have dominated the academic discourse on resilience. But organizations that have an effective safety management system and workplace protections because of the accident hazards intrinsic to their operations are likely to be more attentive to disruptive events that could arise from internal conditions, especially because they will be held accountable for worker safety and offsite impacts on the public [5]. Notorious accidents at the Chevron Richmond refinery and at BP onshore and offshore facilities are clear examples of catastrophic events that arose from internal causes, including top management neglect, middle management negligence, and worker and contractor error.[1]

Regulation also shapes the approach to resilience. Although there is no broad legal mandate that a company or other private organization make itself resilient, or measure and certify its resilience, this does not mean that resilience is merely left to company discretion. Many local and national regulations, building codes, standards and permit requirements apply to the design, siting and operation of facilities in order to protect public health, safety and environmental quality from harms that may arise from foreseeable types of disruptive events. Similarly, workplace safety regulations, and common law doctrines that impose liability for harms due to a company's negligence, have the effect of promoting organizational resilience [6]. Thus, regulatory compliance and liability avoidance contribute to resilience.

Those aspects of resilience that are not mandated by law are left to organizational discretion, as may be the case for installing a cyber-defense or backup energy system, for example. Generally, it can be expected that such matters, collectively or separately,

---

[1]US Chemical Safety Board reports on these accidents are available at https://www.csb.gov.

will undergo a review process regarding their technical and financial feasibility, costs and benefits, value for improving competitiveness and fulfilling contractual obligations, and overall acceptability to top management.

But this review process will also consider insurance as a less-costly alternative to resilience initiatives. A proposed resilience initiative may be rejected when it can be shown that casualty, liability, business interruption and other types of insurance coverage are available, affordable and adequate to cover the estimated losses that would be caused by the type of disruptive event being addressed. According to some policy analysts, this situation obstructs progress towards a safer society because: "insurance regimes reinforce exposure and vulnerability through underwriting a return to the *status quo* rather than enabling adaptive behavior" [7].

A final point for this brief commentary is that improving resilience can be facilitated by recognizing and gaining value from inter-organizational dependencies [8]: for example, by creating effective plans for preventing and responding to disruptive events with the following entities:

- Public infrastructure entities that provide essential services such as water supply, transport and electricity.
- Community and state departments that provide services for emergency response and communications, rescue, evacuation, and medical needs.
- Neighboring industries whose disruption by a major accident would have spill-over disruptive impacts on others.
- Stakeholders and local organizations that can provide public support for measures that prevent risk and cope with consequences, such as by facilitating governmental approval of plans for reconstruction and restoration.

By outreach and discussion of mutual concerns and interests, an organization can develop these dependencies into resilience-improving relationships.

5.2 Public Infrastructure

Modern society needs infrastructures that serve essential public needs for energy, water supply, food, waste disposal, transportation, communications, and protection against natural hazards and activities that endanger health, property and the environment [9]. The foregoing discussion of resilience sets the stage for now considering its meaning and application in the infrastructure context.

Context matters. In the company context, resilience must serve the firm's self interest. In the infrastructure context, it must serve public needs expressed in processes that govern public infrastructures. This is because public-serving infrastructures are mandated by legislation, designed, built, managed and operated by government agencies or public-private partnerships, and funded by the general public, subsets of users, and investors [10].

Another difference is that resilience in the company context is usually seen as an approach to be taken for the purpose of avoiding financial loss and its value will depend on whether insurance or other loss control measures can produce equivalent results at less cost or less need to change from doing business as usual [7]. But in the infrastructure context, resilience as an objective is unquestioned because it fosters initiatives that protect an essential public service.

The societal value of resilience also differs according to context. Disruption of a company's operations will usually have less harmful consequences for the host community than disruption of an essential infrastructure that may cause a cascade of impacts throughout the community. For example, interruption of a public water supply system may disrupt hospitals, health services, households, human consumption, commercial and industrial activities, schools, and other infrastructures that have water-dependent components such as the regional energy system whose power plants need cooling water. Threats to human health and safety will also require rescue and relief initiatives and connections to any alternate water sources until the water supply infrastructure is restored [11].

Thus, modern society needs to ensure the reliability and continuity of operations of its essential infrastructures despite threats posed by natural hazards, industrial hazards, human error and malicious behavior. Over the last decade, such threats and concerns about their impacts have increased as cyberattacks, terrorism, catastrophic accidents, and extreme weather events attributed to climate change have materialized. As a result, public authorities and advisory groups have come to recognize that the stability and security of a community, indeed its resilience, requires improving the resilience of these essential systems.[2]

Improving the resilience of an essential system to ensure its functional continuity or reliability involves a broad range of initiatives: making its physical and managerial components more robust and capable of resisting the likely impacts of foreseeable events, adding backup energy and other supports, developing the ability to isolate or bypass critical points whose failure would cause system-wide failure and having redundant features in place to replace their functions, enhancing monitoring and maintenance, planning to ensure that alternate services are readily available, and preparing for emergency response, rescue, relief for those who are distressed, and quick repair and restoration of service, for example [12]. As discussed earlier, a coherent approach to resilience will also involve cooperative relationships with other infrastructures, agencies that provide emergency response and relief functions, and stakeholders and local groups who can facilitate implementation of many of the foregoing initiatives.

But the resilience-improving process is complex. It often involves local, state and national levels of government, each with its own priorities and constraints, for approvals and financing, and dealing with competing interest groups and diverse stakeholders. It may also encounter pressures to expand or adjust the infrastructure because of dynamic conditions such as population expansion or dispersion, more

---

[2]Critical Infrastructure Sector Resilience Reports, U.S. Department of Homeland Security. As of July 2017, reports on 16 sectors have been published: e.g. chemicals, communications, food and agriculture, energy, financial services, water and wastewater systems, transportation. Available from https://www.dhs.gov.

stringent performance standards set by regulators as in the case of water quality, new zoning and environmental constraints, and cultural change in the areas being served such as the transition from an industrial or agricultural community to an upscale residential community. And resilience-improving may be confronted by advocates for those who will feel disregarded by traditional "top down" crisis management approaches to resilience, or those who have been underserved because of historic discrimination.

In the U.S., the federal Department of Homeland Security has taken a lead role in defining the resilience challenges and best practices for several types of essential public infrastructure, such as a public water supply system [13]. To move beyond these generalities, it is instructive to briefly review ongoing efforts by the management of an actual infrastructure to improve its resilience and maintain continuity of operations: the Massachusetts Water Resource Authority (MWRA). This public authority was created to manage and operate a vast public water supply infrastructure to serve eastern Massachusetts, a densely populated region that includes several million residents, several thousand companies, and numerous service firms, universities, hospitals and research facilities.[3]

MWRA draws water from 2 reservoirs in rural central Massachusetts, the Quabbin and Wachusett. The reservoirs are surrounded by watersheds comprised of forested and sparsely populated lands that are state-owned or otherwise state-controlled to prevent developments and activities that would contaminate the water held in the reservoirs or damage the watersheds that replenish the reservoirs with fresh water of high quality. Over 200 million gallons per day are drawn and transported over 100 miles through deeply-buried pipelines and rock tunnels to a treatment facility and thence into a network of pipelines and tunnels that serve Boston and 50 other cities and towns.

The water is tested throughout the system, treated using ozone and UV light, and chlorine when necessary, does not require filtration, and is considered the best in the US for its natural quality and absence of contaminants when delivered to the communities. The communities being served are purchasers of the water and each uses its own pipeline network to bring the water to its ultimate users.

According to MWRA, its objectives are to provide reliable, uninterrupted delivery of water that meets all applicable water quality standards for human consumption, and to have the resilience capacity to prevent and respond to system breakdowns. Knowing that its system is tightly-coupled, it follows an approach adopted by water suppliers and public agencies that calls for identification and evaluation of "single points of failure" that could render the system unable to meet its design basis, and the development of redundancies, controls and security measures to eliminate the single points of failure when possible, or protect them when they cannot be eliminated for technical, economic or other reasons.

---

[3] Discussion of the MWRA system is based on numerous public reports and other public documents available at the MWRA website and the website of its Water Supply Citizens Advisory Committee (WSCAC): https://www.mwra.com and https://www.mwra.com/02org/html/wscac.htm.

This approach has been brought to bear on the main transmission system and involves redundancy projects that create operational flexibilities: for example, construction of redundant tunnels, pipelines, new interconnections and replacement of antiquated and untested control equipment of uncertain functionality with new controls that allow parts of the system to be taken off-line for regular inspection and repair without system shutdown. The large scale projects require substantial capital investment, long term financing, approvals by several tiers of public officials, and coordination with communities and other infrastructures.

The quest for resilience throughout the system also involves many other initiatives. These include physical barriers and law enforcement for protecting key assets, improving the robustness of facilities, and various safety management measures in order to prevent contamination of the water supply and degradation of the watersheds by human activities and natural hazards, e.g.:

- Security measures to prevent access by terrorists or trespassers that include barriers, surveillance, and coordination with police and others.
- Watershed restrictions on construction and installations of fuel and chemical storage tanks, waste disposal, and septic systems.
- Additional watershed and reservoir restrictions that limit public access, boating, camping, mountain biking and other recreational activities.
- Monitoring and actions to prevent invasive aquatic species and invasive plants and insects in order to protect the storage and transmission systems and forested watersheds.
- "Environmental policing" to prevent contamination by wildlife and birds.
- Contingency and emergency response plans, including simulations, for containment of spills from nearby rail and road accidents.
- Maintenance of dams and other fixtures that enable water impoundment, and spillways to deal with excessive stormwater.
- Preparations to draw from alternate sources of water and to carry out repairs as needed.

As this example shows, the resilience-improving agenda for an essential public infrastructure must have redundancy and reliability projects that enhance the capacity to prevent system shutdown. It must also encompass protective measures and barriers that protect assets and enhance robustness and resistance to potentially-disruptive impacts throughout the system. Implementation requires coordination with other units of government that have expertise and resources.

5.3 Public Policy

As the foregoing discussion indicates, the concept of resilience for an essential public infrastructure has been expanded to encompass many sub-concepts such as resistance, reliability, redundancy, robustness, rescue, relief, restoration and recovery. This thematic aggregation provides a formula for maintaining or returning to the *status quo*, and thereby can cause disregard for adaptive management and use of new technologies when addressing threats of disruption. This situation is reinforced when the infrastructure involves a network of major facilities and the dependency of other infrastructures, such as an energy infrastructure comprised of power plants, transmission lines, pipelines, fuel storage tanks, equipment, and interconnections that are vital for other infrastructures.

The implementation of a resilience agenda for a public infrastructure such as transportation or water supply now includes security-enhancing activities that are exempt from requirements for transparency and public involvement. These activities are designed to prevent intentional malicious acts such as cyberattacks or use of explosives or chemicals and require secrecy to be effective. They usually derive from national security mandates and templates, and involve "top down" command and control management. Their inclusion in infrastructure resilience is necessary but needs to be compartmentalized to prevent infecting other resilience-improving actions with secrecy.

Finally, there is the challenge of ensuring that resilience-improving activities, especially those needed to maintain and restore the *status quo*, are also consistent with and reinforce responsible approaches for infrastructure sustainability [14]. These approaches to sustainability may be focused on addressing components of an infrastructure that are major polluters and contaminators of community air and water, inefficient consumers of limited resources, cause major accidents, or fail to promote conservation and prudent use of the energy, water or other services they provide. Thus, the opportunity is presented to fashion resilience projects that also heed and advance such sustainability initiatives to the extent feasible.

5.4 Conclusion

Resilience is a worthy objective for any organization. For an essential public infrastructure, it is a necessary objective that takes on meaning and societal value in accordance with the task at hand. Most often, the task is strategic preparedness for foreseeable types of disruptive events and their cascading consequences. As modern society becomes more complex, its stability is increasingly dependent on the performance of public infrastructures that serve essential needs, are tightly coupled, and function in a complex web of dependencies and interdependencies. Thus, ensuring the resilience of such infrastructures is a challenging core function of a society that strives to effectively govern risk and achieve its own resilience.

References

1. J. Xu, L. Xue, On resilience based risk governance, in *Resource Guide on Risk Governance*, International Risk Governance Council (2016)

2. Y. Haimes, Strategic preparedness for recovery from catastrophic risks to communities and infrastructure systems of systems. Risk Anal. **32**(11) (2012)
3. M. Bruneau, The 4 R's of resilience and multi-hazards engineering: the meta-concept of resilience (2007)
4. J. Fiksel, I. Goodman, A. Hecht, Resilience: navigating towards a sustainable future. Solutions **5**(5), 38–47 (2014)
5. A.R. Hale, M.S. Baram (eds.), *Safety Management — The Challenge of Change* (Pergamon, Oxford, 1998)
6. E. Marsden, Risk regulation, liability and insurance: literature review of their influence on safety management. Cahier de la sécurité industrielle 2014-08. Foundation for an Industrial Safety Culture, Toulouse, France (2014)
7. P. O'Hare, I. White, A. Connelly, Insurance as maladaptation: Resilience and the business as usual paradox. Environ. Plan. C: Polit. Space **34**(6), 1175–1193 (2016)
8. NAP, Linking private and public infrastructure interests, in *Disaster Resilience: A National Imperative* (National Academies Press, Washington, 2012), pp. 127–128
9. MASS, Key infrastructure, in *Masachusetts Climate Change Adaptation Report* (Commonwealth of Masachusetts, 2011), pp. 53–70
10. World Bank, Main financing mechanisms for infrastructure projects (2017)
11. S.E. Chang, Infrastructure resilience to disasters, in *Frontiers of Engineering* (National Academies Press, Washington, 2010), pp. 125–134
12. T.D. O'Rourke, Critical infrastructure, interdependencies, and resilience. Bridge **37**(1), 22–29 (2007)
13. DHS, Sector resilience report: water and wastewater systems, Technical report, U.S. Department of Homeland Security, Office of Cyber and Infrastructure Analysis (2014)
14. B. Walker, Resilience, adaptability and transformability in socio-ecological systems. Ecol. Soc. **9**(2) (2004)

Chapter 6
Human Performance, Levels of Service and System Resilience

Miltos Kyriakidis and Vinh N. Dang

Abstract The concept of resilience has spread widely in recent years and is broadly used to examine the dynamic response of critical sectors to disruptions. Resilience is frequently associated with the ability of a system to return to normal operational conditions subsequent to a shock event. Numerous definitions of resilience have been introduced and measures of resilience developed. Yet, the existing literature shows a lack of agreement in operationalising resilience. This chapter expresses resilience in relation to systems performance and levels of service. As people at all levels of an organisation play a significant role on creating (or not) resilience, the human contribution to the resilience of critical infrastructure is discussed. Here, the four resilience cornerstones, i.e., knowing what to do, look for, expect, and has happened, help structure the discussion. This standpoint is found to support a robust operationalisation of resilience.

Keywords Human performance · Critical infrastructure · Levels of Service Resilience cornerstones

6.1 Introduction

Over the last decade, the concept of resilience has developed substantially [1]. The literature (e.g., [2–5]) comprises diverse definitions, resulting in the lack of a universal understanding of the construct [4] and in turn its further operationalisation [6, p. 2713]. Consequently, work is still required to make the notion comprehensible and usable for the relevant stakeholders [7].

M. Kyriakidis (✉)
ETH Zurich, Future Resilient Systems, Singapore-ETH Centre,
Singapore, Singapore
e-mail: miltos.kyriakidis@frs.ethz.ch

V. N. Dang
Laboratory for Energy Systems Analysis, Energy Divisions, Paul Scherrer Institute,
Villigen, Switzerland
e-mail: vinh.dang@psi.ch

© The Author(s) 2019
S. Wiig and B. Fahlbruch (eds.), *Exploring Resilience*, SpringerBriefs
in Safety Management, https://doi.org/10.1007/978-3-030-03189-3_6

To this aim, this chapter focuses on the operational dimension of resilience. Its scope is threefold: first, to associate resilience with a systems levels of service; second, to investigate how this could be implemented and used by relevant stakeholders in their daily operations; and, third, to investigate the relationship and contribution of humans to resilience considering that, in order to cope with real world complexity, individuals as well as organisations constantly adjust their performance to the current conditions.

6.2   Resilience as System Behaviour and Service Levels

Resilience is described as the operational behaviour of a system subsequent to an endogenous or exogenous shock event [8], and it is associated with four response behaviours. The first response behaviour, namely robust, illustrates a system that can fully recover after a shock event. The second and third response behaviours, i.e., ductile and collapsing respectively, refer to a system that can either recover its basic and critical functions or collapse after a shock event. Finally, the fourth response behaviour, adaptive behaviour, represents a system that could reach a performance level higher than the original level, e.g., when the system is reconfigured during its recovery and restoration.

Additionally, adopting the studies by Robert et al. [9] and UC Quake Centre [10], five levels of service for a system are identified, as follows:

- Optimal level of service (OpLoS): the theoretical condition for which the system was planned and designed.
- Normal level of service (NLoS): The system is performing as required and expected, achieving its mission to supply the anticipated level of service, while all the systems outputs are in their normal state.
- Acceptable level of service (ALoS): The systems performance is partially degraded, with one or more of the systems outputs in a disturbed mode. Still, due to the action(s) taken, i.e., contingency plan(s), the system can maintain the service quality at acceptable levels and limit its degradation.
- Unacceptable level of service (ULoS): The systems performance is severely degraded and despite the action(s) taken its degradation has become unacceptable. The system is no longer able to accomplish its mission.
- Out of service level (OLoS): Discontinuation of the service.

The above classification does not provide exact thresholds to determine when the performance of a system changes from one LoS to another. Such thresholds are determined and described by the organisations according to their expectations, requirements and needs. In the case of rapid transit or metro systems, for instance, the LoS comprises: (i) in-service on-time performance (quality), (ii) the frequency of

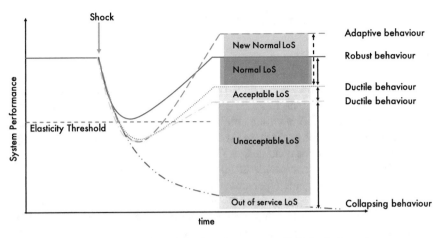

Fig. 6.1 The association of resilience behavioural responses and the LoS of a system

service (quantity) and the trains headways, the (iii) average load factor (quantity).[1] Figure 6.1 shows how the LoS above can be associated to system behaviours. First, the NLoS region lies between the elastic response corresponding to robust behavior and some ductile behavior with longer-term degradation. This implies that "normal" does not imply operating the system continuously at 100% and that some margin remains available. The ALoS region lies just below this region; some post-disruption LoS degradation, below normal, typically remains tolerable. Third, there is an ULoS where the system continues to operate but does not meet reduced post-disruption expectations. This region is bounded by the OLoS, which results from the inability to maintain any service due to a collapsing behavior. In contrast to the LoS named above, the "new normal" LoS results from an adaptive behavior that enables an increase in performance.

Resilience is applicable to safety-critical as well as other systems. For the former, whose loss or failure has direct implications, resilience emphasizes continued and correct operation in the wake of disruptions [11]. For other systems, whose purpose is not a safety function but considered essential (critical) infrastructure, resilience implies continued operation as well. Naturally, if this service must be safe, it suggests continued operation and excludes operation with reduced safety levels, at least in this discussion. Nevertheless, the degradation or loss of such service may have indirect safety implications. In the case of public transport systems as discussed in this chapter, for instance, overcrowding on station platforms may have safety consequences or the resulting congestion on other transport modes may hinder emergency services as a knock-on effect.

---

[1]Service Performance Indicators used by the Los Angeles County Metropolitan Transportation Authority, https://www.metro.net/about/metro-service-changes/service-performance-indicators/.

6.3   The Human Contribution to Resilience

Woods describes [12] resilience as *a parameter of a system that captures how well that system can adapt to handle events that challenge the boundary conditions for its operation.* Such events may occur due to (i) limitations in the plans and procedures, (ii) the tendency of systems to adapt given changing pressures and expectations for performance, and (iii) environmental changes. The systems response capacity to challenging events lies partially in the expertise, strategies, and tools that people employ to respond to certain challenges [12].

It is therefore clear that people at all levels of an organization, e.g., frontline personnel, middle management personnel, and top policy decision makers, are able to create (or not) resilience by adjusting their performance to current operational conditions [13]. Research [14] has already defined four fundamental cornerstones that describe a resilient organisation and are associated with the human contribution to resilience:

- Knowing what to *do*, which refers to the ability of responding to regular and irregular disruptions and disturbances by adjusting normal functioning or activating readymade responses.
- Knowing what to *look for*, which refers to the ability of monitoring that which is or could become a threat in the near term. The monitoring shall cover both what happens in the environment, and what happens in the system itself, i.e., its own performance.
- Knowing what to *expect*, which refers to the ability of anticipating developments and threats further into the future.
- Knowing what *has happened*, which refers to the ability of learning from experience.

Reviewing the cornerstones, a continuous loop of interactions can be observed, as shown in Fig. 6.2, where human involvement is divided into two main levels. The first level, in the upper half of the figure, refers to the contribution of the frontline personnel as well as the responses of the crisis teams, including management. This level of involvement includes the short-term actions/tasks of the personnel, and represents those individuals, or teams within an organisation who respond after the occurrence of a disruption and who react and manage to recover the LoS. Depending on the type of disruption, there may also be opportunities to limit the magnitude of degraded service with a possibly consequential positive impact on its duration. All of these actions are associated with the *what to do* and *look for* cornerstones.

The second level, in the lower half of the figure, refers to a longer-term organisational response across the whole spectrum of an operation, including any normal and unexpected situations. It is assumed that the organisations knowledge of *what to do* and *what to look for*, on which the response to a disruption is built, is itself built upon the organisation previous experience and anticipation. Experience is derived from the organisations learning from past events, while anticipation refers to its ability to identify potential, future threats. Learning and anticipating, in other words, cor-

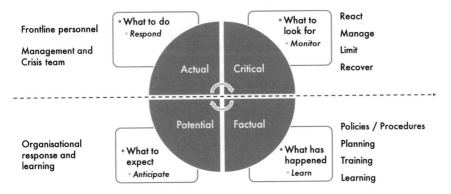

Fig. 6.2 The resilience capabilities loop as function of human contribution (adopted from [13] and extended by the authors)

responding to the *what has happened* and *what to expect* cornerstones respectively, together form the basis for preparedness, which is transformed concretely into the plans, policies, procedures and training that are applied in the actual response to a disruption.

6.4 Resilience Operationalisation Using the Four Cornerstones and the LoS Concept

This section demonstrates the operationalisation of resilience using the four resilient cornerstones and the concept of service levels in the transportation sector. Data was collected from publicly available reports [15, 16] that describe two major disruptions of the Singaporean metro system that occurred in December 2011 within a period of two days.

The first disruption, on December 15, lasted five hours and affected about 127,000 commuters. A second disruption occurred on December 17, spanned over seven hours and affected some 94,000 commuters. An investigation found that both disruptions were preventable and caused by a combination of factors, none of which individually would have resulted in the disruptions. The official investigation report [15] describes the events leading to the disruption as follows:

> The immediate cause of the stalling of the trains was damage to their Current Collector Device (CCD) "shoes" due to sagging of the "third rail", which supplies electrical power to the trains. During both incidents sections of the third rail sagged after multiple "claws", which hold up the third rail above the trackbed, were dislodged. With their CCDs damaged, the trains were unable to draw electricity from the third rail to power their propulsion and other systems such as cabin lighting and air-conditioning.

The investigators found that the December 15 incident

was initiated by a defective fastener in the Third Rail Support Assembly (TRSA), which damaged the Current Collector Device (CCD) shoes of the trains that passed the incident site. In the process, these trains destabilised the third rail system elsewhere along the network, and the forces generated by the CCD shoes of multiple trains impacting the sagging third rail caused three more claws at the incident site to be dislodged, such that the third rail came to rest on the trackbed. Thereafter, this segment of the third rail became totally impassable to all trains.

The second incident

was triggered by one or more "rogue trains" which suffered not easily detectable CCD shoe damage when passing the 15 December 2011 incident site as the third rail was progressively sagging. In its haste to resume revenue service on 16 December 2011, the metro personnel did not conduct a sufficiently thorough investigation, such that the CCD shoe damage on the rogue train(s) went undetected. Had the investigation been thorough, the incident on 17 December 2011 might have been prevented.

In addition, the analysis of the events prior to the disruptions in [15] identified numerous factors that contributed to the incidents, such as:

- Defects on train wheels that resulted in severe vibration.
- Gauge fouling, or contact with the third rail system by passing trains due to the separation between the third rail and the running rail.
- Design of the current third rail claw.
- Shortcomings in the maintenance work culture within Singaporean Mass Rapid Transit (SMRT).
- Shortcomings in the maintenance and monitoring regime, mainly in the context of ageing assets.

This example highlights a service disruption due to a combination of failures. Had the failures happened independently they would have not produce any substantial disruption on the system. Using the four resilient cornerstones, it could be claimed that all contributing factors are primarily associated with the SMRTs inability to provide its employees with the appropriate means (e.g., policies and procedures) to execute their tasks.

Regarding the levels of service, the SMRT managed to restore its service in timely manner, while also providing alternative travel options to its customers (e.g., replacement buses). In spite of the preparedness to manage the disruptions indicated by this response, the LoS in both disruptions were deemed unacceptable, as implied by the fine imposed by the Singaporean Land Transport Authority [16]. Thus, the elasticity threshold should not be related to acceptability, while the LoS, if measured as a momentary or average capacity, is not sufficient per se to discuss service degradation. Instead, a service degradation measure needs to consider the duration (width) of the disruption and not only its magnitude (depth).

With respect to resilience cornerstones, the SMRT seemed to have learnt from the experience of the first disruption; and managed better the incident related to the second disruption. Replacement buses and alternative travel options were deployed. Considering the longer duration of the second disruption, here it appears important to determine the boundaries of a systems LoS and service degradation. Indeed, the

SMRT duration of disruption was longer, yet the passengers were better served and transferred to their destinations. Hence, a broader measure of the overall performance of the system (or the measure of service degradation) considers not only the customers whose metro was not available but also completed passenger trips independent of mode, e.g. replacement services.

The SMRT example shows that organisations shall not only focus on preparing and planning how to handle with individual shock events, but also to account for potential consequential effects and their impact on the systems overall resilience. This example underscores the importance of ensuring that recovery is not only timely but also durable, i.e., placing the system into an "as good as new" state in reliability terms. The incident on the December 17 may have been prevented if the SMRT was not in haste to resume its service on December 16, subsequent to the disruption on December 15. Such haste led to the deployment of not sufficiently investigated trains and in turn to the second disruption in the system.

6.5   Conclusions

Resilience is broadly used to study and understand the response of critical sectors to disruptions. However, the operationalisation of the notion has not sufficiently been explored. In this chapter, resilience was presented in association with a systems performance and its degradation in terms of service levels as an undesired outcome distinct from those related to potential hazards to the public and environment. Resilience was described as the operational behaviour of a system subsequent to the occurrence of an endogenous or exogenous shock event. Further, five levels of service for a system were identified, i.e., the new normal, normal, acceptable, unacceptable and out of service level. The association between service levels and system resilience was also shown. Moreover, the acceptability of the systems response (service level trajectory) can be seen as largely unconnected with whether this response is elastic. Ultimately, the resilience of systems that deliver a service is defined not in terms of whether the system response exhibits an elastic behaviour but rather whether the service level trajectory is acceptable. Specifically, a ductile or inelastic response with a longer-term service level degradation may be acceptable; the acceptability criteria for the system will instead be based on response criteria such as the minimum service level maintained during the peak of the disruption, the magnitude of the longer-term degradation, and an overall service loss that combines the duration and magnitude of degraded service.

People at all levels of an organisation play a significant role in creating (or not) resilience. This chapter examined the human contribution to resilience, whereby the four resilience cornerstones clearly provide a helpful lens. Yet, it could be seen that the functions the cornerstones describe need to be interpreted on two levels. First, on the organizational level in terms of anticipating threats, learning from disruptions, and incorporating the lessons thereof into contingency plans and training. Second, for the frontline personnel at the "sharp end", the functions become responding to a disruption per the procedures, monitoring whether the actions taken are successful to prevent and mitigate service degradation or recover service, and anticipating the systems evolution to enable a proactive response.

This chapter does not provide any figures of merit about a systems resilience involving the LoS or the probability/frequency and duration of the service degradation. Thus, future research will focus on evaluating different systems and their preparedness against unexpected events, while it will also identify human critical tasks and scenarios that could lead to significant losses.

Acknowledgements This work was conducted at the Future Resilient Systems at the Singapore - ETH Centre, which was established collaboratively between ETH Zurich and Singapore's National Research Foundation (FI 370074011) under its Campus for Research Excellence and Technological Enterprise programme.

References

1. D.D. Woods, Four concepts for resilience and the implications for the future of resilience engineering. Reliab. Eng. Syst. Saf. **141**, 59 (2015)
2. R. Bhamra, S. Dani, K. Burnard, Resilience: the concept, a literature review and future directions. Int. J. Prod. Res. **49**(18), 5375–5393 (2011)
3. S.B. Manyena, The concept of resilience revisited. Disasters **30**(4), 434–450 (2006)
4. H. Zhou, J. Wang, J. Wan, H. Jia, Resilience to natural hazards: a geographic perspective. Nat. Hazards **53**(1), 21–41 (2010)
5. T.B. Sheridan, Risk, human error, and system resilience: fundamental ideas. Hum. Factors **50**(3), 418–426 (2008)
6. D.E. Alexander, Resilience and disaster risk reduction: an etymological journey. Nat. Hazards Earth Syst. Sci. **13**, 2707–2716 (2013)
7. S. Hosseini, K. Barker, J.E. Ramirez-Marquez, A review of definitions and measures of system resilience. Reliab. Eng. Syst. Saf. **145**, 47–61 (2016)
8. Future resilient systems. Technical report, Singapore-ETH Centre (2015)
9. B. Robert, W. Pinel, J.-Y. Pairet, B. Rey, C. Coeugnard, Y. Hmond, *Organizational resilience - concepts and evaluation method* (Presses Internationales Polytechnique, 2010)
10. Levels of service performance measures for the seismic resilience of three waters network delivery. Technical report, UC Quake Centre, University of Canterbury, New Zealand (2015)
11. N. Leveson, N. Dulac, D. Zipkin, J. Cutcher-Gershenfeld, J. Carroll, B. Barrett, Engineering resilience into safety-critical systems, in *Resilience Engineering: Concepts and Precepts*, ed. by E. Hollnagel, D.D. Woods, N. Leveson (Ashgate, Aldershot, 2012), pp. 95–124
12. D.D. Woods, Resilience engineering: Redefining the culture of safety and risk management. Hum. Factors Ergon. Soc. Bull. **49**(12), 1–3 (2006)
13. S.W. Dekker, E. Hollnagel, D.D. Woods, R. Cook, Resilience engineering: new directions for measuring and maintaining safety in complex systems. Technical report, Lund University School of Aviation (2008)
14. E. Hollnagel, Epilogue: RAG - the resilience analysis grid, in *Resilience Engineering in Practice: A Guidebook*, ed. by E. Hollnagel, J. Paris, D.D. Woods, J. Wreathall (Ashgate, Aldershot, 2011)
15. Singapore Ministry of Transport, Report of the committee of inquiry into the disruption of MRT train services on 15 and 17 Dec 2011 (2012)
16. Singapore Land Transport Authority, SMRT to be fined $2 million for december 2011 train service disruptions along the north south line (2012)

Chapter 7
Precursor Resilience in Practice – An Organizational Response to Weak Signals

Kenneth Pettersen Gould

Abstract This chapter looks at resilience from the descriptions of organizational strategies and practices in a regional airline operating regular commercial flights at short runway airports. Like many organizations facing environmental changes and intensive operational demands, the airline faces cascades of disturbances and friction in putting plans into place, requiring the ability to extend performance. This study demonstrates that different types of resilience exist and that precursor resilience is more about the organizational expansion of expectancies than individuals or groups managing the unexpected. This clarification adds depth to the understanding of resilience in aviation and similar organizational contexts, and the chapter takes issue in discussing how resilience varies and is different according to level in organizations or systems, place, time, resources, and competencies. This extends ongoing research efforts identifying specific types of resilience and their requirements based on a closer grounding of the concept in empirical studies.

Keywords Precursor resilience · Weak signals · Organization · High reliability Management

7.1 Introduction

Resilience is seen by many as an answer to organizational survival in a more complex and uncertain world [1–9]. Previous work to develop a theory on organizational resilience has anchored resilience in two suggested beliefs [8]. First, resilient organizations possess an "intelligent wariness" [10]. They treat successes lightly and are leery of the potential of the unexpected [11]. Second, resilient organizations strive for operational perfection under chronic unease. They operate under the belief that they are imperfect but can over time learn through events and near events [8].

While adding to our understanding of modern organizations, resilience as theory has become highly generalized and abstracted [12]. The identification of what makes

K. Pettersen Gould (✉)
University of Stavanger, Stavanger, Norway
e-mail: kenneth.a.pettersen@uis.no

resilience in organizations has, to a limited degree, been clarified by theoretical development and individual empirical cases [9, 12, 13]. Organizational resilience has been viewed as the result of beliefs in organizations as well as emotional, behavioral, and cognitive processes that enable organizations to cope successfully with and learn from unexpected events [8].

Previous researchers argued that empirical evidence gathered from studying organizations running high-risk systems, such as nuclear power and aviation, suggested that resilience has quite different forms in organizations and quite different—if not contradictory—requirements [12, 14]. Resilience varies and differs according to levels in organizations or systems, time, resources, and competencies. These different types of resilience include *precursor resilience*, which relates to monitoring and keeping operations within a bandwidth of conditions and acting to restore these conditions quickly as a way of managing risk. Previous research has shown that the accumulation of small interruptions can compromise the safety of a system just as readily as a larger event [8, 15]. Important in this respect is the possibility that resilient organizations notice relevant weak signals more quickly. In addition, *restoration resilience* consists of rapid actions to resume operations after a disruption whereas *recovery resilience* puts damaged systems back together to establish a new normal that is at least as reliable and robust as before—if not improved. Previous research has provided more material on how personnel and cognitive challenges associated with each differ [14]. Preparation and training for restoration resilience, for example, may diminish attention to prior structures and competencies for precursor resilience.

The failure to pare out and empirically ground the concept of resilience into different types has led to misleading perceptions of the concept. Its generalized treatment may have also discouraged the development of more specific findings of use to organizations themselves in promoting various kinds of resilience [16]. These generalizations are particularly problematic in terms of the importance of resilience as a topic on the safety research agenda, such as our understanding of organizations that must carefully manage high-risk systems that—if mismanaged—could lead to catastrophic failures and cost many lives. Previously, the combination of discourses on complexity and resilience has led to distorted depictions of high reliability organizations (HROs) [12]. In HROs, the generalized promotion of resilience threatens to undermine our ability to distinguish localized adaptations to unpredictable situations from conditions where localized adaptations actually become a negative development in relation to the pursuit of larger reliability and safety goals.

This chapter offers some additional insights into precursor resilience. A key issue is the organizational, meso-level strategies for resilience relied on within the airline, which include other resources than the individual/group level accounts of resilience often promoted within the safety domain [17]. In fact, the precursor resilience we witnessed took advantage of elaborate structures of planning, organizational factors, and competencies that are often critiqued as anticipatory planning [11] and bureaucratic approaches to safety [18] working against resilience. Our findings suggest a relationship between anticipation and certain types of resilience in HROs; furthermore, structures of planning do not have to stand in the way of successful adaptation in high-risk systems. In addition, what a quick and rapid response constitutes in precursor resilience should be viewed in relation to the type of weak signals to which actions respond. In relation to the time needed to analyze and respond, these events are also different than the more serious disruptions or accidents.

The chapter is written based on the study of a regional airline operating regular commercial flights at short runway airports. These airports have runways between 800 and 1500 m and with few systems for instrument-based approach, landing, and takeoff. The study included data collection over several periods in 2013, consisting of interviews with airline managers and safety management personnel. In addition, two research stays were completed at the airline headquarters in 2013 and 2014, as well as visits to two airline base stations.

7.2   An Organizational Strategy for Precursor Resilience

The airline operated scheduled short runway flights under smaller safety margins, yet with reliability standards and a safety performance equivalent to commercial civil aviation in general [12]. Responding to societal demands for service regularity on a network of 26 short runway airports servicing Norway's most remote coastal regions, the airline developed a specialized strategy for high reliability to fit with societal demands. The airline had a strategy to deal with high input variability related to challenging topography, diverse infrastructure, and changing weather conditions. These conditions required a higher degree of pilot judgment compared to commercial civil aviation in general. The role of pilot judgment could well lead to actions away from accepted standards by individual pilots under pressure to provide service to otherwise isolated rural communities or by pilot temperament to accept higher risks as part of their self-confidence in their own skills [12]. In this unusual setting and in the face of flight conditions one would not think acceptable within the context of HROs, the airline took advantage of precursor resilience and kept operations within a bandwidth of conditions.

As summarized by [17], the common use of the resilience concept relates to the ability of an organization or a system to return to its normal condition or functioning after an event has disturbed its regular state. Thus, the resilience literature often refers to dynamic capabilities, adaptive capacities, and performance variations as key topics. Broadly, there is no order in the application of resilience as it is seen to be related to unplanned, unpredictable, and largely undirectable aspects of emergent properties of complex systems [19, 20]. The resilience identified in this study is different from the "rebound from failure" resilience or the process of "managing the unexpected" described in earlier literature discussing HRO research [11]. By using strategies for system monitoring and the analysis of interruptions and departures from baseline performance, the airline was able to take into account uncertainties and act on identified early warnings. These strategies included the careful and continuous monitoring of flight operations, in relation to both the airline's internal operations and environment. This acting to keep or restore operations within a bandwidth of conditions could involve a network of internal and external actors, demonstrating that resilience in the context of high reliability can be structured and require coordinated alterations of action.

7.3 Precursor Resilience in Practice: An Example

Operating well-understood aircraft technology and having elaborate systems for planning, the airline shared key conditions for high reliability with commercial civil aviation in general. The organization relied on an extensive framework of analytic and experiential knowledge detailed in maps, formalized flight limitations, and procedures—in many cases, specific for each of the short runway airports. Keeping operations uniform to their level of reliability meant flying in and out of airports where precursor conditions could be specific to the individual airport and current flight conditions. However, as seen in earlier studies of HROs, the formalization of tasks did not support the centralization of authority [21]. In fact, much of the formalization related to documenting and reinforcing the elaborate organizing that being resilient required. In the domain of safety, for example, a Safety Service Office (SSO) was available for advice and guidance on safety-related matters to all nominated safety personnel across levels and departments. Organized by the airline's safety manager, the SSO monitored the performance of management systems in the area of safety and was responsible for the delivery of safety services to the other departments in the organization. In addition to the SSO, a number of cross-departmental groups and functions were available. For example, a Safety and Compliance Review Board (SCRB) headed by the CEO was responsible for interactions between safety and compliance, as well as other major issues of flight safety connected to operations. A local safety management group (SCAG) for flight operations worked across procedures, practices, and people. In addition, the airline chose to establish a health and safety advisory group (HSAG) meeting. According to the airline's safety manager, the idea was to have representatives from frontline personnel look into their areas of operation, together with safety coordinators and the safety manager. The mandate for the HSAG was to evaluate past events and practices in order to identify lessons learned, while also making proactive plans to avoid reoccurrences. The group gave HSE-strategic advice to SCRB and advised the local safety management groups (i.e., SCAG and FSAG) on HSE-related issues that should be considered in their action plan, ultimately issuing recommendations. The HSE advisory group also acted as a working group for SCRB on rising safety issues and could be asked to give detailed information on such issues. One example of precursor resilience emerged in the work processes of the flight data monitoring group (FDMG). FDMG regularly conducted overviews and analyses of flight data monitoring (FDM) data. FDM data was gathered from across flight operations, including all takeoffs and landings, on a routine basis. It was mandatory for the airline to report FDM discrepancies of a serious nature, but the information could also provide systemic insights into even small changes in relation to established operational limitations, quality, and reliability criteria. In the work of the FDMG, we found cases where the group had identified systemic departures from operational limitations and initiated processes within the airline to respond to these early warnings.

The FDMG routinized a form of watchfulness, which was a quality nurtured by the airline's safety management. In relation to safety management practices within the airline, watchfulness can be described as the continuous monitoring, analysis, and questioning of one's knowledge of operations at the many different airports operated by the airline and the discrete risks that flying to and from them may present. Short runway operations required a sensitivity to operations [11], a recognition of diversity, and an attention to detail. Although most of the airline's management personnel had many years of experience from operations in the cockpit, maintenance, or ground handling, they did not rely on existing patterns of action as being sufficient for reliability and safety.

A full process of precursor resilience, where operations were restored within bandwidth boundaries, was identified related to the analysis of a relatively high number of unstabilized approaches and excessive bank on approach. These incidents were individually not serious events, in the sense of representing an accident risk. However, viewed as a pattern in the analysis, they were interpreted as early warnings at several of the short runway airports.

The frequency of these events raised concerns that there may be precursor conditions causing the trends. During interviews, we were informed of how the safety manager and the FDMG had engaged in a process of abductive analysis [22], creating a hypothesis of what could be causing the trends. At three airport's, incident trends were associated with pilots repeatedly adapting their landing approaches to a combination of technological changes and constraints in the airports infrastructure and support systems. In other words, the planes were getting bigger and flying faster than the existing airport infrastructure and systems were designed for. As technology and infrastructure are systemic issues involving the infrastructure owner and civil aviation authority, the airline itself had only limited influence on restoring conditions within acceptable bandwidths, and a quick response required a shared understanding of the risk and response across organizations. By sharing the analysis with other stakeholders and communicating risk, the systemic issues were agreed upon as precursor conditions causing the airlines pilots to adapt. Because of this process, investments in new technology and changes in airport inflight procedures were made at three short runway airports.

Following these changes, the airline experienced a 26–40% decrease in incident trends. An interesting illustration from this process is a picture of FDM data imported into Google Earth maps, providing a rich description of systemic aircraft movements (see Fig. 7.1). In the figure, each yellow and red triangle represents an excessive bank on approach. Using such maps in consultations with the owner of the airport infrastructure and the national aviation authorities, the airline could communicate their analysis of risk based on a richer graphical representation, not just numbers in a table.

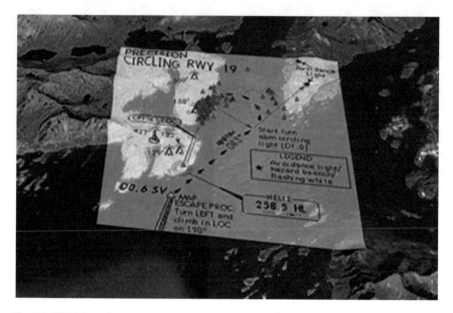

Fig. 7.1 FDM data of excessive bank events at airport

7.4 Concluding Remarks

We recognize that precursor resilience was an addition to other features of high relia-bility in the airline. These include acting based on extensive and detailed procedural systems. This is comparable to the role of resilience described in other research on high reliability management and HROs [12, 14, 21–23]. During the interviews, air-line management representatives underscored the importance of a successful merger between best practice with planning and procedures. However, in this study, the mar-gins or bandwidths of operations that people had to accept were more differentiated than we have seen in civil aviation and perhaps other HROs in general [23].

The type of resilience described in this paper is covered by Aaron Wildavsky's broad definition of being resilient as vitally prepared for adversities that requires "improvement in overall capability, i.e., a generalized capacity to investigate, to learn, and to act, without knowing in advance, what one will be called to act upon" [9]. Yet within this broad definition, resilience has to be treated differently. Our research indicated that promoting precursor resilience relies on strategies at organizational and system levels—in this case, including structures for collecting system-wide data, planning, and coordinated action. Although resilience in general is an ability to respond quickly, the events to which precursor resilience responds are not major disruptions or accidents. This gives an organization more time to act, as weak signals require time to analyze as well as coordinate responses. The latter is of particular importance within the context of high-risk systems, where localized adaptions can become negative development in relation to larger reliability and safety goals. In fact, in association with precursor resilience, organizational strategies and structures seem to be a prerequisite for detecting both early warnings and responsive capacities to act when high reliability is key for the organization.

This research supports the idea that resilient organizations, through their updated and nuanced picture of ongoing operations, are able to "[…] parlay that understanding into more targeted and timely investments in tolls or actions that can defuse emerging vulnerabilities and risks before harm results" [11]. The type of resilience identified here, in association with high reliability, is also of a specific type [12] and different from more generalized accounts of resilience in HROs [11]. In addition, it is important to note that the precursor resilience we have described is safety oriented and related to a specific, but not limited, set of events and early warnings, which are defined as relevant by their association to the risks the airline has as their key concerns (i.e., a serious event or accident with an aircraft). Thus, no claims can be made that the airline as an HRO can identify and restore all types of adversities [24], nor can we claim that the resilience we have documented can be retained or provide protection against the rigidness and dangers of proceduralization.

Acknowledgements This research was supported by the Norwegian Research council TRANSIKK program REALTRANS grant. I would like to thank Paul Schulman for insightful discussions.

References

1. A. Boin, M. Van Eeten, The resilient organization. Public Manag. Rev. **15**(3), 429–445 (2013)
2. G. Grote, *Management of Uncertainty. Theory and Application in the Design of Systems and Organizations* (Springer, Berlin, 2009)
3. E. Hollnagel, C.P. Nemeth, S.W. Dekker (eds.), *Resilience Engineering Perspectives: Remaining Sensitive to the Possibility of Failure* (Ashgate, Aldershot, 2008)
4. J.M. Kendra, T. Wachtendorf, Elements of resilience after the World Trade Center disaster. Disasters **27**(1), 37–53 (2003)
5. T.R. La Porte, *Organized Social Complexity: Challenge to Politics and Policy* (Princeton University Press, Princeton, 2015)
6. Y. Sheffi, *The Resilient Enterprise* (MIT Press, Cambridge, 2005)
7. K. Sutcliffe, T. Vogus, Organizing for resilience, in *Positive Organizational Scholarship*, ed. by K.S. Cameron, J.E. Dutton, R.E. Quinn (Berrett-Koehler, San Francisco, 2003)
8. M.M. Waldrop, *Complexity Theory: The Emerging Science at the Edge of Order and Chaos* (Simon and Schuster, New York, 1992)
9. A. Wildavsky, *Searching for Safety* (Transaction Books, New Brunswick, 1988)
10. J. Reason, *Managing The Risks of Organizational Accidents* (Ashgate, Aldershot, 1997)
11. K.E. Weick, K.M. Sutcliffe, *Managing the Unexpected: Assuring High Performance in An Age of Uncertainty* (Jossey-Bass, San Francisco, 2001)
12. K.A. Pettersen, P.R. Schulman, Drift, adaptation, resilience and reliability: toward an empirical clarification. Saf. Sci. (2016). (in press)
13. M.D. Bruijne, A. Boin, M. van Eeten, Resilience: exploring the concept and its meanings, in *Designing Resilience: Preparing for Extreme Events*, ed. by L.K. Comfort, A. Boin, C.C. Demchak (University of Pittsburgh Press, Pittsburgh, 2010), pp. 13–32
14. E. Roe, P.R. Schulman, *Reliability and Risk: The Challenge of Managing Interconnected Infrastructures* (Stanford University Press, Stanford, 2016)

15. J.W. Rudolph, N.P. Repenning, Disaster dynamics: understanding the role of quantity in organizational collapse. Adm. Sci. Q. **47**(1), 130 (2002)
16. D.D. Woods, Four concepts for resilience and the implications for the future of resilience engineering. Reliab. Eng. Syst. Saf. **141**, 59 (2015)
17. S. Wiig, B. Fahlbruch, in *Exploring Resilience: Proposal for a NeTWork Workshop* (2017)
18. S.W. Dekker, The bureaucratization of safety. Saf. Sci. **70**, 348–357 (2014)
19. J. Bergstrm, R. van Winsen, E. Henriqson, On the rationale of resilience in the domain of safety: a literature review. Reliab. Eng. Syst. Saf. **141**, 131–141 (2015)
20. T.J. Vogus, K.M. Sutcliffe, Organizational resilience: towards a theory and research agenda, in *Proceedings of the 2007 IEEE International Conference on Systems, Man and Cybernetics* (IEEE, 2007), pp. 3418–3422
21. P.R. Schulman, The negotiated order of organizational reliability. Adm. Soc. **25**(3), 353–372 (1993)
22. K.A. Pettersen, Acknowledging the role of abductive thinking: a way out of proceduralization for safety management and oversight? in *Trapping Safety into Rules How Desirable or Avoidable is Proceduralization?* ed. by C. Bieder, M. Bourrier (Ashgate, Farnham, 2013)
23. E. Roe, P.R. Schulman, *High Reliability Management: Operating on the Edge* (Stanford University Press, Stanford, 2008)
24. K.A. Pettersen, Understanding uncertainty: thinking through in relation to high-risk technologies, in *Routledge Handbook of Risk Studies*, ed. by A. Burgess, A. Alemanno, J.O. Zinn (Routledge, London, 2016)

Chapter 8
Leadership in Resilient Organizations

Gudela Grote

Abstract This chapter focuses on organizations' ability to change between different modes of operation as a key adaptive capacity that fosters resilience. Four modes are described which represent responses to low versus high demands on stability and flexibility respectively. The operational requirements for leaders both in enacting the different modes of operation and in instigating switches between the modes are detailed. Strategic recommendations are outlined that should help organizations to build the needed leadership abilities and to support organizational change towards better handling fundamental tensions and trade-offs embedded in the requirement to stay in control while facing unexpected uncertainties.

Keywords Operational leadership · Strategic leadership · Stability · Flexibility Team adaptivity · Organizational culture

8.1 Introduction

Resilience has been defined in simple terms as a system's ability to "bounce back" after disturbances, and, through learning from those situations, to "bounce forward" and increase the system's adaptive capacity for handling surprises [1, 2], thus incorporating reactive and proactive responses to uncertainty [3]. Much of what has been written about resilience aims to describe general characteristics of organizations which enable resilience, such as the necessity to continuously monitor, anticipate, respond, and learn [4] or to manage trade-offs in the face of challenged system boundaries, which Woods [2] has termed graceful extensibility. Woods also discusses the need to shift between performance regimes [5], which can be traced back to the early studies on high reliability organizations where these organizations' ability to switch

G. Grote (✉)
ETH Zürich, Zurich, Switzerland
e-mail: ggrote@ethz.ch

© The Author(s) 2019
S. Wiig and B. Fahlbruch (eds.), *Exploring Resilience*, SpringerBriefs
in Safety Management, https://doi.org/10.1007/978-3-030-03189-3_8

between different modes of operation in response to changing demands has been identified as a crucial source of resilience [6, 7].

The necessity to manage trade-offs and tensions stemming from dynamic and possibly conflicting requirements is echoed in organization theory and strategic management. Managing paradox, for instance by enabling simultaneous exploitation and exploration, routine and innovation, or stability and flexibility, is discussed as a core organizational capability [8]. In the following, I propose to conceptualize resilience within such a general organizational framework. Drawing on the rich research traditions in the organization and management sciences permits to anchor our understanding of how organizations manage risk and uncertainty in validated constructs for organizational design and leadership.

After a brief review of research on safety leadership, I will discuss how leaders in high-risk systems have to cope with simultaneous stability and flexibility demands as one of the most fundamental tensions underlying organizational resilience, which stems from the requirement to control systems while staying responsive to uncertainty. I will argue that leaders play a key role in helping individuals and teams to address these complex demands on adaptive behavior, while they themselves have to also build and employ a portfolio of leadership styles matching different situations. Beyond these operational leadership requirements, strategic leadership is needed to establish supporting structures in the design of the organization such as standards and cultural norms and to foster organizational change towards building adaptive capacity at every level of the organization. Finally, some practical recommendations are given for training leaders and for promoting an appreciation for different worldviews which is necessary to increase acceptance of the tensions inherent in the different modes of operation as cornerstones of resilience. Given the scarcity of literature on strategic safety leadership to date, this chapter provides important insights for practitioners who strive to design resilient organizations along with building the necessary leadership capabilities, but it is also a rich source for future research aimed at bridging organization theory and safety science.

8.2 Research on Safety Leadership

As in the leadership literature more generally [9], much of the research on safety leadership has been concerned with identifying effective styles of operational leadership. Especially transformational leadership—that is leadership aimed at motivating employees through inspiration and charisma—has generally been found to be positively related to safety outcomes. However, in a recent meta-analysis, Clarke [10] has shown that transactional leadership—the counterpart of transformational leadership aimed at an exchange of rewards for fulfilling expectations—supports both safety participation and compliance, while transformational leadership supports mainly safety participation. According to the findings by Zohar and Luria [11] transformational leadership is particularly important for promoting safe behavior when the priority of safety is not sufficiently embedded in company values. The requirement to adapt leadership styles to situational requirements more generally has been demonstrated by Yun, Faraj and Sims [12] who found that directive leadership in medical emergency teams was more successful in complex cases and with less experienced team members, while an empowering style was effective in less complex cases and with more experienced teams.

Another approach to leadership is to define it in functional terms as exerting influence on others in order to determine and achieve objectives. The tasks and processes involved in leadership are emphasized rather than the formal leadership role. This perspective also provides the foundation for the concept of shared leadership, which argues that leadership functions can be fulfilled not only by the formal leader, but possibly by any team member [13]. In high-risk teams shared leadership has been shown to promote safety, for instance in shock trauma teams [14] and anesthesia teams [15]. The situation becomes even more complex, when several teams interact as so-called multi-team systems, as is the case for cockpit crews and cabin crews. Bienefeld and Grote [16] analyzed shared leadership within and across cockpit and cabin crews. Behavior observations of 84 aircrews handling a simulated emergency (smoke of unknown origin in the cabin) showed that overall successful aircrews — which achieved a safe landing with all passengers adequately protected — were characterized by more shared leadership in the cabin, but not in the cockpit. More leadership by the captain was related to team goal attainment, that is a safe landing, independently of whether the aircrew overall achieved its goal. Furthermore, more leadership of pursers, that is, the formal leaders of the cabin crew, towards the cockpit crew was evident in successful aircrews. The authors discuss their findings in view of the pursers' crucial role as boundary spanners for achieving overall success and the support for this role through shared leadership in the cabin crew. More generally, the study indicates that some caution is needed in judging the benefits of shared leadership. It appears that shared leadership needs to be carefully balanced with leadership by the formal leader.

Finally, there may be situations where personal leadership itself becomes less important due to substitutes being in place such as standards that prescribe work processes in great detail or very experienced employees who know what to do without a leader telling them. In cockpit crews, it was indeed found that better performance in a simulated non-routine situation (landing an aircraft without the flaps and slats on the wings working) was linked to coordination patterns with little leadership and much implicit coordination in highly standardized work phases (take-off and landing) and more leadership during the less standardized work phase of preparing for the unusual landing [17]. Moreover, substitution may affect different leadership functions differently. For instance, in anesthesia teams consisting of more experienced nurses and less experienced, but formally responsible residents (i.e. physicians training at a hospital to become a specialist in a particular field of medicine), successful performance was linked to residents being mostly involved with information search and structuring, while nurses took over problem solving [14].

This brief review shows that there are many contingencies at play that will influence whether certain leadership responses will foster safe operations. These contingencies will now be explored further based on a typology of situations crafted around the fundamental tension between demands for stability and flexibility in organizations. Subsequently, specific requirements for operational and strategic leadership will be distilled from this typology and discussed with respect to building resilient teams and organizations.

8.3 Stability and Flexibility Demands in Organizations

Grote, Kolbe and Waller [18] have proposed to distinguish different situations teams may face in organizations along the two dimensions of stability and flexibility demands. Stability demands arise from organizational requirements for predictability, reliability, and efficiency [19]. These demands are created within organizations to ensure managerial control and maximum productivity, but they may also stem from external sources such as regulatory bodies whose task it is to keep organizational functioning within certain bounds [20]. Perrow [21] has stressed technology as another source of stability demands, specifically tight coupling of work processes with few buffers and little fault tolerance. For teams, these demands entail the need to employ the same work processes, for instance to foster traceability of decisions, and to produce the same outcomes consistently and reliably. Flexibility demands, on the other hand, result from the necessity or desire to widen the scope of action and to innovate [19]. Usually, highly dynamic and uncertain environments are stressed as the main source of flexibility demands. However, flexibility demands also arise from within the organization due to complex production processes or possibly the opposite—highly routinized work processes, where over-routinization and complacency are to be avoided by introducing variation and change [22]. In any of these situations, teams are expected to be responsive to changing demands based on variability of their behavior, possibly even by proactively creating new work processes and outcomes.

Aiming to match stability and flexibility demands with appropriate coordination mechanisms, Grote et al. [18] have relied on substantial organizational research conducted over many decades. Most fundamentally, this research has shown that structural coordination mechanisms are better suited to respond to stability demands, while personal coordination mechanisms help to create flexibility [23]. This led them to hypothesize the following four situations with corresponding coordination patterns.

1. Experiential situations: When both stability and flexibility demands are low, as for instance in team debriefings where the focus is on sharing knowledge and learning outside of acute work pressures, coordination mostly happens among team members without much reliance on formal leadership or organizational rules.
2. Exploitation situations: When stability demands are high and flexibility demands low, as in many process control tasks, the emphasis is on efficient production, usually enabled by structural coordination mechanisms embedded in technology

and standard operating procedures, leaving little need for leadership or mutual adjustment among team members.

3. Exploration situations: When stability demands are low and flexibility demands high, for instance in teams that have to innovate at all cost, coordination happens by mutual adjustment and shared leadership to bring all team members' competences and resources to bear on idea generation and implementation.

4. Ambidextrous situations: When stability and flexibility demands are high because both highly reliable performance of complex tasks and fast reactions to unpredictable change are required, a broad range of coordination mechanisms has to be employed in parallel, helping teams to maintain control, e.g. through directive leadership and/or strong shared norms, and be adaptive, e.g. through sharing leadership tasks.

Teams may have to move quickly between the four conditions and switch their mode of operation accordingly. A surgical team may perform a routine operation (high stability, low flexibility) followed by a complex emergency operation (high stability, high flexibility). It will also undertake team debriefings (low stability, low flexibility) and may engage in experimenting with a new operating technology (low stability, high flexibility). As a consequence, continuous monitoring of stability and flexibility requirements and of necessary adaptations is crucial. To date research has mostly addressed adaptation in operational teams, however top management teams are also confronted with varying requirements for stability and flexibility, for instance when having to ensure effective production processes during major organizational change or when moving the organization towards abandoning old business models. The leadership requirements for enabling adaptation in any kind of team and for building adaptive capacity and resilience in the organization as a whole are discussed next.

8.4 Leadership for Resilience

From the proposed perspective of enabling resilience through adaptive switches between modes of operation, three fundamental leadership requirements can be derived. The first one is the leaders' ability to be adaptive themselves, that is to transform their own role and behaviors according to the stability and flexibility demands their teams face. Their behavior repertoire has to stretch from fostering stability through rules and personal direction to sharing leadership responsibility and giving up control when high flexibility is required to becoming just another team member during informal learning and knowledge exchange. Furthermore, they have to be capable of sensing changes in demands and to prepare themselves and the team for the appropriate switches between modes of operation. The corresponding competencies and skills leaders should possess have been described in terms of cognitive and behavioral complexity [24] and more recently as paradox-savvy leadership [25]. Core is a leader's ability to perceive, understand, and proactively address tensions such as maintaining control while letting go of control and maintaining continuity while simultaneously pursuing change [26]. This is not only demanding due to the different behaviors required, but also due to the need to reconcile different world-views embedded in different approaches to stability and flexibility [27, 28].

The second requirement is to design organizational mechanisms that support individual and team adaptivity. Foremost, this concerns structures and standards put in place, which usually are meant to promote stability. Great care has to be taken, though, that the stability created does not lead to rigidity. In the case of rules, for instance, one should also consider to include rules that enable flexibility. These could be goal or process rules, which only specify certain overarching goals or priorities (e.g., "Safety First") or processes to follow in order to decide on the best course of action [29]. An example of a process rule is the "10 for 10" principle that requires a 10 second time-out to plan the next 10 min of coordination during emergencies [30]. Such rules provide stability for team functioning by ensuring that certain processes are adhered to, but support flexibility as well because reflection and speaking up are promoted. Similarly, mandatory debriefings foster both structure and freedom to challenge and adapt existing procedures [31].

The third requirement relates to leaders' role in establishing organizational culture. Beyond building the mindful or informed culture that is generally considered a foundation for resilience [32, 33], the fundamental role of culture as a powerful stabilizing force that helps to coordinate action and integrate work processes in decentralized and flexible modes of operations should be taken into account and employed wisely [34]. As Weick [35, p. 124] has described it: "(Culture) creates a homogeneous set of assumptions and decision premises which, when they are invoked on a local and decentralized basis, preserve coordination and centralization. Most important, when centralization occurs via decision premises and assumptions, compliance occurs without surveillance."

For instance, a shared norm of always speaking up with concerns and ideas will better help mastering unexpected challenges than any attempt to cover all possible turns situations can take by means of standard operation procedures [28]. Regarding the particular nature of cultures that are beneficial for resilience one crucial aspect is respect for the viability of different perspectives on problems and their solutions. Such a culture of interdisciplinary appreciation is at the heart of bringing all knowledge in organizations to bear on finding the most effective ways to promote safety [36, 37] and to adequately address the ensuing paradoxical tensions [26].

8.5   Three Strategic Recommendations

If one accepts my argument that organizational resilience is closely tied to the ability to function in different modes of operation and to successfully manage switches between these modes of operation, three strategic recommendations follow. First, leaders and their teams need to be trained in these abilities. Crew Resource Management training in aviation is a very successful example for such trainings. But even these trainings still have their challenges, such as extending them from cockpit crews to cabin crews along with the necessary appreciation for adaptive delegation of leadership [16]. Amalberti [27] has also pointed to the difficulty of trainings being attached to a particular mode of operation and underlying assumptions and rules for that mode. Thus, if trainings are to enable leaders and their teams to switch been modes operations, these underlying assumptions need to be addressed as well. An interesting example in this respect is a recent study by Weiss and colleagues [31]. The authors showed that after-event reviews conducted as part of training sessions for anesthesia teams led to more speaking up if assertiveness was emphasized, but also to more hierarchy-attenuating beliefs. Only if leaders share the view that hierarchy can come in the way of safety in certain situations, successful transfer of trained behaviors to the real world will ensue.

This leads to the second recommendation. To truly embrace different modes of operation requires bridging the worldviews embedded in the different approaches to managing stability and flexibility. Resilience depends on a shared understanding across professional boundaries of the legitimacy of different kinds of leadership in response to tensions concerning control and adaptive capacity, which may even entail deliberate promotion of uncertainty in some cases [28, 38]. Perspective taking and cross-learning among the different professions involved in safety are crucial to reflect on and reconcile the diverse belief systems and to create shared views on problems and on ways to solve them. Leaders are called upon to facilitate these processes and to encourage the needed organizational change.

The third recommendation, finally, concerns the relationship between operating organizations and regulatory agencies. Different worldviews do not only exist within organizations but - most likely even more so - between organizations, especially when they have very different tasks and roles such as operating versus regulating and inspecting safety-critical processes. Thus, a shared view of the legitimacy of different modes of operation has to also be established between operating organizations and their regulators and auditors. Depending on the given regulatory regime [20] this is a more or less daunting task. When regulation is prescriptive, a shared perspective on what is adequate behavior is more important, but especially acceptance of empowered modes of operation may be more difficult to achieve due to that same regulatory preference for highly standardized processes. Only if an open dialogue between operator and regulator is established, can the operational flexibility which lies at the heart of resilience be effectively realized.

References

1. L. Ilmola, Organizational resilience how do you know if your organization is resilient or not? Resource Guide on Resilience, EPFL International Risk Governance Center (2016)
2. D.D. Woods, Four concepts for resilience and the implications for the future of resilience engineering. Reliab. Eng. Syst. Saf. **141**, 59 (2015)
3. K.A. Pettersen, P.R. Schulman, Drift, adaptation, resilience and reliability: toward an empirical clarification. Saf. Sci. (2016) (in press)

4. E. Hollnagel, J. Paris, J. Wreathall (eds.), *Resilience Engineering in Practice: A Guidebook* (Ashgate, Farnham, 2011)
5. D.D. Woods, Resilience as graceful extensibility to overcome brittleness, IRGC Resource Guide on Resilience, EPFL International Risk Governance Center (2016)
6. T.R. La Porte, P. Consolini, Working in practice but not in theory: theoretical challenges of high reliability organizations. J. Public Adm. Res. Theory 1, 19–47 (1991)
7. K.E. Weick, Collapse of sensemaking in organizations: the Mann Gulch disaster. Adm. Sci. Q. 38(4), 628–652 (1993)
8. J. Schad, M.W. Lewis, S. Raisch, W.K. Smith, Paradox research in management science: looking back to move forward. Acad. Manag. Ann. 10, 5–64 (2016)
9. B.J. Avolio, F.O. Walumbwa, T.J. Weber, Leadership: current theories, research, and future directions. Annu. Rev. Psychol. 60, 421–449 (2009)
10. S. Clarke, Safety leadership: a meta-analytic review of transformational and transactional leadership styles as antecedents of safety behaviours. J. Occup. Organ. Psychol. 86(1), 22–49 (2013)
11. D. Zohar, G. Luria, Group leaders as gatekeepers: testing safety climate variations across levels of analysis. Appl. Psychol. Int. Rev. 59(4), 647–673 (2010)
12. S. Yun, S. Faraj, H.P. Sims, Contingent leadership and effectiveness of trauma resuscitation teams. J. Appl. Psychol. 90(6), 1288–1296 (2005)
13. C.L. Pearce, J.A. Conger, *Shared Leadership: Reframing the How's and Why's of Leadership* (Sage, Thousand Oaks, 2003)
14. K.J. Klein, J.C. Ziegert, A.P. Knight, Y. Xiao, Dynamic delegation: shared, hierarchical, and deindividualized leadership in extreme action teams. Adm. Sci. Q. 51(4), 590–621 (2006)
15. B. Künzle, E. Zala-Mezö, J. Wacker, M. Kolbe, G. Grote, Leadership in anaesthesia teams: the most effective leadership is shared. Qual. Saf. Health Care 19(6), 1–6 (2010)
16. N. Bienefeld, G. Grote, Shared leadership in multiteam systems: how cockpit and cabin crews lead each other to safety. Hum. Factors 56(2), 270–286 (2014)
17. G. Grote, M. Kolbe, E. Zala-Mezö, N. Bienefeld-Seall, B. Künzle, Adaptive coordination and heedfulness make better cockpit crews. Ergonomics 52(2), 211–228 (2010)
18. G. Grote, M. Kolbe, M.J. Waller, The dual nature of adaptive coordination in teams: Balancing demands for flexibility and stability. Organ. Psychol. Rev. (Published online first). https://doi.org/10.1177/2041386618790112
19. J.D. Thompson, *Organizations in Action: Social Science Bases of Administrative Theory* (McGraw-Hill, Toronto, 1967)
20. P.J. May, Regulatory regimes and accountability. Regul. Gov. 1(1), 826 (2007)
21. C. Perrow, A framework for the comparative analysis of organizations. Am. Sociol. Rev. 32(2), 194–208 (1967)
22. C.J.G. Gersick, J.R. Hackman, Habitual routines in task-performing groups. Organ. Behav. Hum. Decis. Process. 47(1), 6597 (1990)
23. A.H. Van de Ven, M. Ganco, C.R. Hinings, Returning to the frontier of contingency theory of organizational and institutional designs. Acad. Manag. Ann. 7(1), 393–440 (2013)
24. D.R. Denison, R. Hooijberg, R.E. Quinn, Paradox and performance: toward a theory of behavioral complexity in managerial leadership. Organ. Sci. 6(5), 524–540 (1995)
25. D.A. Waldman, D.E. Bowen, Learning to be a paradox-savvy leader. Acad. Manag. Perspect. 30(3), 316–327 (2016)
26. T. Reiman, C. Rollenhagen, E. Pietikinen, J. Heikkil, Principles of adaptive management in complex safety critical organizations. Saf. Sci. 71, 8092 (2015)
27. R. Amalberti, *Navigating Safety: Necessary Compromises and Trade-Offs - Theory and Practice* (Springer, Berlin, 2013)
28. G. Grote, Promoting safety by increasing uncertainty - implications for risk management. Saf. Sci. 71, 71–79 (2015)
29. G. Grote, J.C. Weichbrodt, H. Günter, E. Zala-Mezö, B. Künzle, Coordination in high-risk organizations: the need for flexible routines. Cogn. Technol. Work. 11(1), 1727 (2009)
30. M. Rall, R.J. Glavin, R. Flin, The 10-seconds-for-10-minutes-principle. Bull. R. Coll. Anaesth. 51, 2614–2616 (2008)

31. M. Weiss, M. Kolbe, G. Grote, D.R. Spahn, B. Grande, Why didn't you say something? Effects of after-event reviews on voice behaviour and hierarchy beliefs in multi-professional action teams. Eur. J. Work. Organ. Psychol. **26**, 66–80 (2017)
32. J. Reason, *Managing the Risks of Organizational Accidents* (Ashgate, Aldershot, 1997)
33. K.E. Weick, K.M. Sutcliffe, *Managing the Unexpected: Assuring High Performance in an Age of Uncertainty* (Jossey-Bass, San Francisco, 2001)
34. G. Grote, Understanding and assessing safety culture through the lens of organizational management of uncertainty. Saf. Sci. **45**(6), 637–652 (2007)
35. K.E. Weick, Organizational culture as a source of high reliability. Calif. Manag. Rev. **29**(2), 112–127 (1987)
36. J.S. Carroll, Organizational learning activities in high-hazard industries: the logics underlying self-analysis. J. Manag. Stud. **35**(6), 699–717 (1998)
37. G. Grote, Social science for safety: steps towards establishing a culture of interdisciplinary appreciation, in *Proceedings of an International Conference on Human and Organizational Aspects of Assuring Nuclear Safety Exploring 30 Years of Safety Culture, Organized by IAEA, Vienna* (2016)
38. M.A. Griffin, J. Cordery, C. Soo, Dynamic safety capability: how organizations proactively change core safety systems. Organ. Psychol. Rev. **6**, 248–272 (2016)

Chapter 9
Modelling the Influence of Safety Management Tools on Resilience

Teemu Reiman and Kaupo Viitanen

Abstract Descriptions of new safety management tools or suggestions for modifying existing tools on the basis of the principles of the Resilience Engineering paradigm are rare. This chapter introduces an evaluation checklist for adaptive safety management that can be used in analyzing the influence of safety management tools on resilience. Three commonly used safety management tools are inspected from the Resilience Engineering perspective to understand how they can be utilized for enhancing resilience in safety-critical organizations. The chapter concludes that the traditional tools of safety management focus heavily on constraining activity, but they do have a positive influence on the system's general adaptive capacity. This effect is often unintentional, but the tools can also be used intentionally for this purpose, which requires becoming aware of both the direct and the indirect effects of the existing methods.

Keywords Safety management · Tools · Resilience · Adaptive management

9.1 Introduction

Resilience Engineering is becoming a widely-recognized safety management paradigm in modern safety science. One of its essential purposes is to propose an alternative, complementary approach to safety, one that acknowledges the importance of variability, decentralized control and the complex and emergent phenomena that result from systemic interactions. However, descriptions of new safety management

T. Reiman (✉)
Lilikoi Consulting, Helsinki, Finland
e-mail: reimanteemu@gmail.com

K. Viitanen
VTT Technical Research Centre of Finland Ltd, Espoo, Finland
e-mail: kaupo.viitanen@vtt.fi

tools or suggestions for modifying the existing tools on the basis of the principles of the Resilience Engineering paradigm are rare. In recent systematic reviews of Resilience Engineering literature, it was found that the development of practical tools have received little attention [1, 2], and in studies where Resilience Engineering has been utilized for devising or modifying safety management tools, the focus has predominantly been on measuring resilience (with a few exceptions of studies that concern training for resilience capabilities) rather than creating or maintaining it [2]. In order for Resilience Engineering to gain ground among the safety professionals working in safety-critical industries, more concrete clarifications of how Resilience Engineering can be applied in the daily work of a safety professional are needed. In this chapter we examine three commonly used safety management tools from the Resilience Engineering perspective to understand how they can be utilized for enhancing resilience in safety-critical organizations.

9.2 Adaptive Safety Management

Resilience Engineering and the so called Safety-II perspective emphasize that system safety is an emergent property of the system and should be seen as the system's ability to succeed under varying conditions [3, 4]. In order to succeed, the organization requires adaptive capacity in addition to standard operating procedures. Organizations need to be able to respond to both expected and unexpected disruptions. They need to change and remain stable at the same time. As a consequence, management of a complex safety-critical organization is an inherently contradictory activity. It requires balancing between various tensions, competing demands and irresolvable dichotomies that can never be completely solved [5]. Sometimes the tools that are used to solve one type of problem can have unintended effects on the system, and even generate other, different kind of problems. At the same time, the heterogeneity of tools can in fact be a necessity for safe activities: for instance, sufficient variety is required (e.g., in terms of interpretations) to regulate the safety-critical activities or facilitate learning [6, 7].

Thus, the Resilience Engineering paradigm implies that different, even opposing tools are needed for managing safety. To be able to proactively manage the development activities of an organization, a model is needed to guide the selection and use of the development tools and methods. We utilize the revised model of adaptive safety management [8] originally developed by Reiman et al. [5].

In the model of adaptive management we have included three tensions Fig. 9.1. The selection of tensions is based on the assumptions that a complex sociotechnical system is multilevel (i.e., involves upper and lower systemic levels), has the ability (and tendency) to self-organize and involves interactions between multiple agents [5]. The first tension, levels of system goals, addresses the questions "why" and "what", and stems from the multilevel nature of the system. This tension also involves the idea of temporality, namely that system goals are longer-term, and local goals are shorter-term. The second and third tensions represent safety management strategies

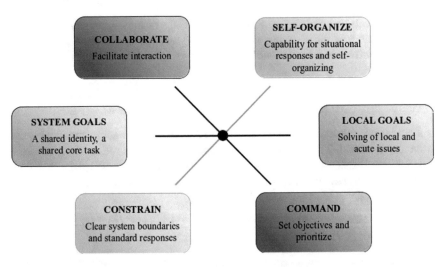

Fig. 9.1 Model of adaptive safety management, adapted from [5, 8]

that aim to manage the self-organization and the interactions between actors. They address the questions of "control" (second tension) and "power" (third tension). Each tension is characterized by contradicting safety management principles (the boxes in Fig. 9.1). In order for safety management to be functional in a sustainable manner, it must be in a state of "dynamic equilibrium": it should have the capability to utilize all of the principles, regardless of whether they are at odds with each other. This means that the safety management tools, and the way in which they are utilized by the safety professionals, should also be sufficiently diverse in order to cover the whole spectrum of the model.

9.3   Evaluation Checklist

The model of adaptive safety management can be utilized for analyzing and selecting practical tools for management of safety from a Resilience Engineering perspective. We demonstrate the use of the model by describing an evaluation checklist and use it to analyze a selection of well-known, traditional safety management tools. The checklist is intended to be used by safety professionals and can be used as a starting point when an organization develops an improvement program or evaluates the effectiveness of its current safety management tools. The purpose of the checklist is to find out whether a given safety management tool supports or hinders the fulfilment of the principles of adaptive safety management and under what condition. The detailed description of each of the adaptive management principles, along with the checklist of questions are described below.

System goals principle refers to shared core tasks of the organization as well as its identity, and how the company sees itself. They are the shared guiding principles according to which decisions should be made, thus steering general adaptive capacity.

Is the tool in line with company strategy, company objectives and top management expectations?

Does the tool help communicate or internalize the organizations shared core task?

Local goals principle refers to the need to pay attention to acute and local issues in the organization. This includes solving specific problems related to subsystems and their functioning. Often solving the local issues does not immediately contribute to system goals, but without solving the local issues, time and resources cannot adequately be devoted to the system goals either.

Does the tool help employees in their daily problems and operational difficulties?

Does the tool help employees to solve acute safety issues?

Collaborate principle refers to facilitating interaction and connections between members of the organization. Connections and interaction between employees at all levels, both horizontally and vertically, are needed in order to guarantee organizational cohesiveness, communication and enough structure for the system to act in a coherent manner and to organize in a decentralized manner when needed. By creating connections between the various actors in the organization, the system also gains situational adaptive capacity due to the possibility of sharing task related information or helping others in their tasks.

Does the tool help employees participate in decision-making processes or design of their own work or the tools they use in their work?

Does the tool create opportunities for discussion between managers and employees?

Does the tool create or serve as an arena of interaction between organizational members?

Command principle refers to setting objectives and prioritizing. Leaders need to select areas where they will focus their effort and to emphasize some connections and some persons over others, depending on their potential contribution to organizational goals. Generally this means that not everyone's wishes can be fulfilled and not everything can be a priority at the same time. In this role, the manager decides what is important and what is not important for the organization.

Does the tool help top management prioritize safety initiatives?

Does the tool give decision-making authority to selected safety professionals or management representatives?

Encourage principle refers to creating capability for situational responses and ability to self-organize activities, structures and mind-sets based on the current or anticipated situational requirements. Encourage also aims to create general adaptive capacity to the organization and requires a reluctance to simplify and an ambition to break up typical categorizations. This means increasing the variance in the system instead of categorizing and (supposedly) decreasing the potential sources of variance, it means giving the organization options for action, cf. [9, 10].

Does the tool facilitate novelty or bring new insights?

Does the tool give permission and preconditions for the personnel to develop and adjust their own rules, roles, and practices?

Does the tool help diversify thinking or doing?

Does the tool facilitate the development of general competences?

Constrain principle refers to striving for clear system boundaries and standardized performance within these boundaries. Roles and responsibilities are a key feature enabling the coordination of activities. For effective coordination, it is important that the expectations concerning working practices are as clear as possible. This requires certain shared decision-making principles that can be embedded in the operating procedures. By creating standard responses the system gains specific adaptive capacity for predefined events and generic capacity due to optimizing of resource use. Organizations also need to set barriers against typical human errors and violations. Constraining can also be ideological, for example if a particular management philosophy is selected and taken as a granted approach to all issues that the organization faces. This makes employee behavior more predictable.

Does the tool use normative, predetermined criteria for performance?

Does the tool set boundary conditions or limits on operations?

Does the tool seek to standardize thinking or doing?

Finally, the following, more general topics are evaluated:

How is the effect of the tool dependent on the ways the organization manages a) system goals – local goals, b) constrain – encourage, c) command – collaborate?

How is the effect of the tool dependent on the way it is implemented?

9.4 Discovering the Adaptive Potential of Safety Management Tools

We selected three well-known safety management tools, safety audits, classroom safety training and reporting systems, to demonstrate how the checklist can be used in practice. The tools were chosen due to the author's previous experience with conducting safety audits [11] and implementing a safety concerns reporting system in the nuclear industry, and carrying out various types of safety training.

Safety auditing is a popular method for evaluating the extent to which a set of predetermined requirements is met. Because of this, they naturally tend to constrain rather than encourage and command rather than collaborate. In principle they can be used to encourage adaptive capacities and self-organizing, if these are taken as criteria when defining the requirements against which performance is audited. Due to the focus on requirements, audits can lead the organization to focus on local goals, that is, those requirements that the audit identifies as not conforming to expectations.

There can also be a tendency to later lose interest when the non-conformity is closed, further decreasing the focus on system goals.

Classroom safety trainings are a commonly used method for safety improvement. For instance, they can develop safety-related awareness, behavior or attitudes [12–14]. Trainings can be targeted at any level of the organization, from shop-floor to top management. Safety trainings typically highlight the importance of safety as a shared value and an organizational priority (system goals). However, the abstract nature of classroom training may hinder the usefulness of the trainings in real work situations.

Various types of reporting systems are utilized for collecting information from the employees. Management uses the feedback received via reporting systems to identify organizational problems and help individuals and teams learn from these mistakes in order to perform better in the future. The usefulness of reporting system is dependent on many conditions, for example, whether the personnel trust that the reports are handled in a just manner, and whether the personnel actually is able to observe issues worth reporting.

Table 9.1 summarizes the various influences these tools exhibit when they are examined in light of the model presented in Fig. 9.1 and the questions listed above.

Table 9.1 The influences of safety auditing, classroom safety training, and reporting on the principles

		Support	Hinder	Condition
Goal	Local	**Auditing** can help identify flaws that the auditee has become accustomed to, which can help the auditee's continuous improvement to achieve its local goals	The abstract and isolated nature of classroom **training** typically does not help in solving daily challenges at work	n/a
	System	**Trainings** highlight the importance of safety as a shared value and facilitating it through acculturation	The way of issuing (and later closing) non-conformities (NC) can lead the auditee to only focus on the issues identified in the **audit**, and losing interest once the NC has been closed	There need to be higher level requirements either from the management system or industry standards that formulate the criteria in the **audits**

(continued)

Table 9.1 (continued)

		Support	Hinder	Condition
Control	Encourage	**Trainings** encourage adaptive capacities by increasing knowledge and skills, and by widening the understanding of what safety means in a given context	**Audits** predominantly focus on verifying compliance and existence of structures instead of actual organizational potential to perform in various situations	In principle **audits** can be used to encourage adaptive capacities and self-organizing, if these are taken as criteria when defining the requirements against which performance is audited
	Constrain	**Auditing** supports constraining by verifying compliance against predefined standards or requirements	**Trainings** aim to provide shared values promoting joint response to expected contingencies	Issues worth **reporting** need defining, but not in too constraining manner to allow the observing of emerging issues
Power	Collaborate	n/a	Due to a power distance between auditee and the auditor, future collaboration can suffer after an **audit**	If there is a lack of trust and collaboration, **reporting** systems are likely to not function properly
	Command	**Auditing** of can be used to steer the auditee to focus on the issues of interest to the auditor	**Reporting** systems can hinder excessive top-down control by giving employees alternative channels to raise concerns	The **auditor** needs enough authority to command the auditee to answer questions and show evidence of issues considered. **Training** is dependent in part on the authority of the trainer to require certain behaviors from the trainees

9.5   Conclusions

This chapter introduced a checklist that can be used in analyzing the influence of safety management tools on resilience. The chapter argued that in management of complex safety-critical systems, several opposing safety management principles need to be taken into account. The question arises, can we, and should we develop tools that contribute equally to all opposing principles? Or should we instead try to optimize the contribution of a selected tool to one principle and seek to remove or control its effects on the other principles? There are probably no clear answers to these questions. It seems, however, that the commonly-used safety management tools have a couple of principles they primarily focus on. This may be due to their historical development in parallel to the models of safety (from technical, human error, organizational factors to resilience). It is important for safety professionals to be aware of how the tools that they use can affect the properties of the system they are trying to manage. The checklist presented in this chapter may help in identifying those effects. Selection and use of safety management tools always includes trade-offs and choices.

For improving resilience, this chapter highlights two points: first, the traditional safety management tools focus heavily on command and constrain, but they do have an influence on the system's general adaptive capacity by either intentionally encouraging it, or unintentionally by hindering it via indirect effects. Second, it is possible to develop tools that modify, maintain or monitor the adaptive capacity of the organization. This can be done by creating a completely new one such as the Functional Resonance Analysis Model [15], but also by modifying an existing tool. Thus, the Resilience Engineering approach to safety management requires not just developing new tools, but rather understanding the influence of current tools and building on the strengths (and weaknesses) of those tools.

References

1. R. Patriarca, J. Bergström, G.D. Gravio, F. Costantinoa, Resilience engineering: current status of the research and future challenges. Saf. Sci. **102**, 79–100 (2018)
2. A.W. Righi, T.A. Saurin, P. Wachs, A systematic literature review of resilience engineering: research areas and a research agenda proposal. Reliab. Eng. Sys. Saf. **141**, 142–152 (2015)
3. E. Hollnagel, Prologue: the scope of resilience engineering, in *Resilience Engineering in Practice: A Guidebook*, ed. by E. Hollnagel, J. Paris, D.D. Woods, J. Wreathall (Ashgate, Farnham, 2011), pp. xxix–xxxix
4. E. Hollnagel, *Safety-I and Safety-II: The Past and Future of Safety Management* (Ashgate, Farnham, 2014)
5. T. Reiman, C. Rollenhagen, E. Pietikäinen, J. Heikkilä, Principles of adaptive management in complex safety–critical organizations. Saf. Sci. **71**, 80–92 (2015)
6. S. Antonsen, Safety culture and the issue of power. Saf. Sci. **47**(2), 183–191 (2009)
7. K.E. Weick, Organizational culture as a source of high reliability. Calif. Manag. Rev. **29**(2), 112–127 (1987)
8. K. Viitanen, T. Reiman, Building an adaptive safety culture in a nuclear construction project – insights to safety practitioners, in *Proceedings of the 7th Resilience Engineering Symposium* (Liège, Belgium, 2017)
9. G. Grote, *Management of Uncertainty. Theory and Application in the Design of Systems and Organizations* (Springer, Berlin, 2009)
10. G. Grote, Safety management in different high-risk domains – all the same? Saf. Sci. **50**(10), 1983–1992 (2012)
11. T. Reiman, K. Viitanen, Safety culture assurance by auditing in a nuclear power plant construction project, in *Proceedings of the 9th International Conference on the Prevention of Accidents at Work* (2018), pp. 161–169
12. J. Harvey, H. Bolam, D. Gregory, G. Erdos, The effectiveness of training to change safety culture and attitudes within a highly regulated environment. Pers. Rev. **30**(6), 615–636 (2001)
13. IAEA, *Safety Culture in Nuclear Installations: Guidance for use in the Enhancement of Safety Culture*. IAEA TECDOC, vol. 1329 (IAEA, Vienna, 2002)
14. J.E. Roughton, J.J. Mercurio, *Developing an Effective Safety Culture: A Leadership Approach* (Butterworth-Heinemann, Boston, 2002)
15. E. Hollnagel, *FRAM, The Functional Resonance Analysis Method: Modelling Complex Socio-Technical Systems* (Ashgate, Farnham, 2012)

Chapter 10
Resilient Characteristics as Described in Empirical Studies on Health Care

Siv Hilde Berg and Karina Aase

Abstract The concept of resilience needs greater empirical clarity. The literature on resilience in health care, published between 2006 and 2016, was reviewed with the aim of describing resilient characteristics in empirical studies. The chapter documents resilient characteristics at the individual, team, management, and organizational level. The characteristics were related to four overall conceptual categories: anticipation, sensemaking, trade-offs and adaptation. Based on empirical accounts resilience is described as a set of cognitive and behavioral strategies of individuals who enact resilience within an organizational context. The characteristics represented should be seen as examples of how resilience is described in the applied health care research, thus informing possible operationalization of resilience.

Keywords Anticipation · Sensemaking · Trade-off · Adaptation
Resilience in health care

10.1 Background

Health care has become a major field of focus for resilience studies accounting for a considerable amount of the empirical literature. However, a common model for operationalization has not been used in the applied research. This may relate to the lack of conceptual clarity. Several diverse definitions of resilience have been proposed

S. H. Berg (✉)
Division of Adult Mental Health, Sandnes DPS, Stavanger University Hospital,
Postveien 181, N-4307 Stavanger, Norway
e-mail: siv.h.berg@uis.no

K. Aase
Centre for Resilience in Healthcare, Faculty of Health Sciences,
University of Stavanger, P.O. Box. 8600 Forus, N-4036 Stavanger, Norway
e-mail: karina.aase@uis.no

© The Author(s) 2019
S. Wiig and B. Fahlbruch (eds.), *Exploring Resilience*, SpringerBriefs
in Safety Management, https://doi.org/10.1007/978-3-030-03189-3_10

over the last decade (e.g. [1, 2]), and researchers argue whether resilience is a unified concept or a compilation of multiple issues [3]. Concept formation is a prerequisite for any attempt to operationalize [4]. Operationalization entails a move from the abstract level to the empirical level, at which the ultimate goal is to find measures that validly and reliably capture the concept under study. The applied use of resilience reflects such action, and synthesizing this knowledge furthers the progress towards conceptualization and operationalization. This chapter therefore aims to synthesize resilient characteristics as described in empirical studies in health care.

We base our chapter on a content analysis of 15 empirical studies of resilience in health care. The literature searches were conducted as part of a larger study (see also [5]). The studies included were peer-reviewed articles or book chapters dated from January 2006 to February 2016. The studies drew from qualitative data within diverse clinical health care settings.

A directed content analysis was conducted [6]. The contents (e.g. resilient actions, attributes, abilities, contingencies, outcomes) of the included studies were collected according to predefined codes at different system levels (individual practitioners, health care teams, management, and organization). Inductive category development was conducted first within each system level. The categories representing resilient characteristics were held at the lowest possible abstraction level, keeping the concepts semantically close to the original findings where possible. Second, category development were conducted across levels to express conceptual categories at a higher abstraction level [7].

10.2   Resilient Characteristics

Resilient characteristics in the 15 included studies were categorized at individual, team, management and organizational level.

10.2.1   Individual Practitioners

Resilient characteristics at the individual level were anticipation, adaptation, sense-making, and cognitive trade-offs.

Anticipation was described as an individual ability for health care professionals to enact resilience. Health care professionals anticipated gaps [8], work demands [9], and traits [10] in the clinical environments and handled each situation before it affected the patient. Pharmacists anticipated intervals of heavier demands and moved some of their work in order to reestablish a margin or buffer of safety to deal with urgent requests [9]. Preconditions for anticipation were related to both individual and situational demands. Ekstedt and Ödegård [8] found that health care professionals anticipated gaps by being sensitive to cues of fragility in the system. Cuvelier and Falzon [10] found that the level of physicians' experience and level of uncertainty affected their ability to anticipate an event in a paediatric anaesthesiology service. When the event overrode the physicians' ability to anticipate the situation, this was either related to physicians' inexperience or experienced physicians facing unknown novel situations.

Adaptations in clinical practice were understood as the result of coping with complexity in terms of unexpected situations, demands, variability attributed to the patient, or new technology [8, 11, 12]. Adaptations were described as an integrative part of daily practice to ensure good outcomes [8, 13]. Brattheim, Faxvaag, and Seim [14] described how variations were anticipated and planned for in vascular surgery. A similar connection was made by Cuvelier and Falzon [10], who found that the anesthesiologists had an anticipatory capacity that enabled them to define an envelope of potential variability before each operation. These findings connect the ability to anticipate and enact adaptations. In everyday clinical practice, the adaptations performed by practitioners were developing rules [11], adapting procedures [15], adding extra consultations and tests [14], conducting a "secret second handover" [16], or taking shortcuts and improvising [8]. Adaptations are performed differently according to the professional's level of competency, roles, and autonomy. Ross et al. [17] found that ward staff provided good outcomes by following treatment protocols to cope with narrow-focused tasks, while specialists could decide to go outside the protocol and take a holistic perspective with higher-level decisions, taking the complexity of each case into account.

Cuvelier and Falzon [10] studied resilient decisions in expected and unexpected events, where anesthesiologists had to adjust to unforeseen variability that required decision making under time constraints and with a high level of uncertainty. Their strategies depended on whether the situation was understood or not. In situations perceived as unexpected, a sense of what was happening was lost (i.e. "collapse of sensemaking" [18]). These findings demonstrate the act of *sensemaking* when facing unexpected events. Without identification of the problem at hand, a protocol could not directly deal with the event, and the anesthesiologists had to make *cognitive trade-offs*, choosing between establishing a correct diagnosis before acting or choosing one of the possible protocols. When the individual understands that an event is changing from normal towards abnormal [19], or towards a crisis [12], adjustments can be made to act proactively to prevent the adverse event or crisis.

10.2.2 Health Care Teams

Resilient characteristics at the team level were categorized as managing trade-offs between competing goals, collaboration across specialists and collective sensemaking.

The ability to manage *trade-offs between competing goals* has been described as a dynamic decision-making process between professionals which collaborates in clinical care. Competing goals and tensions emerge between professionals with different roles, clinical aims, and goals [16, 17]. Tensions are described between the need for safety of the individual patient versus the safety of the patient in the community [16] or between patients' versus nurses' need for safety [15]. Further tensions emerge between professionals due to different goals in care [16]. In order to adapt to these tensions, health care teams make trade-offs based on their experience and decide what to sacrifice [12, 16]. This decision-making process is described as flexible, dynamic (constantly under re-evaluation), and highly context dependent [15, 16]. Teamwork was considered as an adaptive response to ensure safe work to task demands that could not be met alone. Nurses considered that teamwork maximized their physical, cognitive and emotional resources to successfully manage work demands [15]. Paries et al. [12] introduced the concept of "coopetition", a merging of cooperation and competition, and proposed coopetition as a team resilient characteristic in an ICU understood as an ability to manage competing goals. The doctor can decide that a protocol needs to be adapted, but not be in charge of the implementation of the decision, and depend on the nurses performing the duties. Nurses can meet this request with resistance, as they are put in a risky situation by performing outside the protocol. Coopetition was the resilient response to the diversity of interests that emerged between different professionals.

Collaboration across specialists has been described as a resilient characteristic among specialists across disciplines related to anticipation, mitigation, and decision making. Collaborative cross checks comprise a strategy to monitor decision-making to detect erroneous assumptions and actions and prevent errors from happening. At least two people assess the accuracy and validity of others' assumptions and/or actions. The patterns observed in three health care incidents established that collaborative cross checks enhanced system resilience when the incoming fellow had specialized and interdisciplinary knowledge. The cross checks made the process more observable and explicit [20]. Collaboration across specialists was also found to be a resilient characteristic in diabetes care, in which the specialist teams detected problems early and reduced future risks [17]. These findings connect team expertise to the ability to anticipate and act reactively to problems.

Strategies and activities to achieve *collective sensemaking* have been described as resilient characteristics within and between health care teams. The medical visit is an example of a daily activity, which is important for sensemaking within the intensive care team. A collective understanding of the perceived clinical behavior of the patient was built during the medical visit. The shared understanding obtained from a diversity of professionals observations improved the anticipation of future actions to take when faced with clinical complexity [12]. Direct means of communication such as contacting a clinic directly was described as a strategy to obtain a shared common ground between clinics in the diagnostic process of lung and colorectal cancer [21]. Verbal communication was also preferred to ensure shared understanding during handovers in emergency care. Clinicians felt they could not simply rely on documentation. However, in the dialogue health care professionals could highlight important information, discuss their concerns and question information [16].

10.2.3 Management

A few studies described resilient characteristics at the management level, categorized as anticipatory regulation, and crisis management adjustments.

A resilient characteristic found in the ICU was the *anticipatory regulation* performed by the managing assistant nurse who anticipated patient flow and work demands in the ICU and prevented capacity crisis [12]. Miller et al. [22] found that each level of management aimed to provide staffing resources appropriate to the anticipated patient demand. A strategy to avoid decompensation in periods with increased patient demands beyond the anticipated levels was to maintain and develop compensatory buffers consisting of staff that could be called upon in periods of high demand.

Facing unexpected events or a crisis, *crisis management adjustments* from normal situations towards crisis management involve acknowledging the need to shift from one mode to the other. The crisis response depended on whether the ICU was facing a capacity crisis or a complexity crisis. Crisis management in a capacity crisis was characterized by delegation and decentralization relying on the competence and the sensemaking skills of the teams; however, crisis management in a complexity crisis was characterized by mobilization to increase the level of expertise [12]. These findings relate to sensemaking in cases of unexpected events.

10.2.4 Organization

Resilience characteristics at the organizational level were mainly reflected in the discussion part of the extant studies. Resilient characteristics were related to organizational outcomes of resilience and organizational conditions that supported resilient performance.

Although good *organizational outcomes* of resilient performance were described at the department level, this was not always the case at the organizational level when considering the system as a whole beyond the individual components of the organization. The adaptations that are made can make sense locally, but the outcomes are not necessarily successful at a higher level. Resilient performance can ironically lead to brittleness at the organizational level [11, 16, 21]. Laugaland and Aase [23] found that the outcomes of the adjustments imposed by a system reform were perceived as successful from the hospital's perspective, with mixed outcomes from the primary care perspective, and with poor outcomes from the patient's perspective.

Another characteristic described is the vulnerability emerging when the organization relies heavily on resilient expertise and adaptations at the sharp end, leaving the system brittle [8, 9, 17]. By relying on the specialists' expertise, Ross et al. [17] found that the system was threatened by skill erosion among the ward staff. Brattheim et al. [14] and Ekstedt and Ödegard [8] found that resilient performance at the individual level compensated for the lack of resilience in the health care organization and then became "invisible" to managers. Patterson and Wears [9] studied hospital pharmacies, in which the adaptive capacity was exhausted. They found that the system had relied on individuals working at their maximum capacity every day, thereby stretching the system into brittleness and patterns of decompensation. Without the adaptive capacity, the system was not able to respond to unusual demands or a crisis.

Suggestions of how *organizational conditions* (e.g. structural and cultural) can enhance resilience have been presented in the literature. To enhance individuals' capability for awareness in vascular surgery, IT-based process support can be designed to give real-time process information concerning the actual execution and status of the ongoing clinical process [14]. Smith et al. [21] emphasized the importance of supporting the blunt-end administration's awareness of demands and challenges at the sharp end of the system to better anticipate and adapt to problems.

10.2.5 Conceptualization of Resilience in Health Care

The characteristics described in the empirical studies of resilience are interconnected both within and across levels, according to four overall conceptual categories: anticipation, sensemaking, trade-offs, and adaptations. These four conceptual categories are cognitive and behavioral strategies of individuals who enact resilience within an organizational context. The strategies can be studied at individual, team and management level.

Anticipation is an act of looking forward and relates to the future, which enables individuals to enact proactively and prevent adverse events from happening. At the individual level, practitioners anticipate threats, and at the team level, collaboration among specialists implies anticipation. At the management level, anticipation of demands on the system and regulation of these demands to prevent crisis is included.

Sensemaking is the perception of something that is experienced with regard to the current situation. A sense of what is happening is needed to adapt and make trade'offs in both normal practice and in unexpected events. Individual practitioners make sense of unexpected events, team members share information to understand complex cases and managers make sense of unexpected events and crisis to conduct the adjustments needed.

Trade-offs relate to the act of decision-making and is an adaptive response toward the inherent complexity in every day practice. Two types of trade-offs are common: cognitive trade-offs at the individual level and trade-offs between competing goals at the team level.

Adaptations are adjustments made in work practices as a result of coping with complexity. Diverse adaptations are common at the individual level such as improvisations and adjustment of procedures. At the team level, adaptations are made to manage tensions between goals, and at the management level, adaptations concern the shift from normal mode to crisis management.

Resilience are enacted within an organizational context, thus individual strategies are influenced by multiple *organizational conditions* at different system levels. Further, *organizational outcomes* of enacted resilience are evident locally, across department and institutional levels. The perceptions of such outcomes varies depending on the stakeholder and system level.

10.3 Conclusion

Our conceptualization of resilience in health care based on existing empirical accounts represents resilience as a set of cognitive and behavioral strategies of individuals who enact resilience within an organizational context. The conceptualisation adds conceptual clarity in terms of the applied uses of resilience in health care, thus informing possible operationalization of resilience.

Although the setting in this chapter is limited to health care, theoretical generalizations can be made to other industries and settings. Our analysis finds common characteristics across the system levels, i.e. for individual practitioners, within health care teams, and at the management level. These resilient characteristics are expressed as anticipation, sensemaking, trade-offs, and adaptations. Our findings suggest that, although resilience can be considered as a diverse and interconnected concept, it is not necessarily differently expressed at different levels.

Despite the existence of several frameworks for resilience, which represent concepts for resilience at the organizational level, (i.e. "the four cornerstones of resilience" [2]), the included studies did not document any resilient characteristics at this level. Operationalizing and empirically studying resilience at this system level appear to present challenging topics within the research field. As such, the challenge is how to operationalize and study resilience empirically as a system characteristic, not merely expressed as a sum of individuals who enact resilience.

The outcomes described at the organizational level demonstrate that resilience is not necessarily positive for safety considering the system as a whole. This implies the need to address resilience across institutional borders. To better understand the nuances of resilient characteristics more research is needed on how resilience is expressed within and across different system levels, as well as the outcomes of enacted resilience, and the organizational conditions to enhance resilience at multiple system levels.

References

1. A.W. Righi, T.A. Saurin, P. Wachs, A systematic literature review of resilience engineering: research areas and a research agenda proposal. Reliab. Eng. Sys. Saf. **141**, 142–152 (2015)
2. E. Hollnagel, J. Braithwaite, R.L. Wears, Epilogue: how to make health care resilient, in *Resilient Health Care*, ed. by E. Hollnagel, J. Braithwaite, R.L. Wears (Ashgate, Farnham, 2013)

3. D.D. Woods, Four concepts for resilience and the implications for the future of resilience engineering. Reliab. Eng. Sys. Saf. **141**, 5–9 (2015)
4. G. Sartori, Concept misformation in comparative politics, in *Concepts and Method in Social Science: The Tradition of Giovanni Sartori*, ed. by D. Collier, J. Gerring (Routledge, New York, 2008)
5. S.H. Berg, K. Akerjordet, M. Ekstedt, K. Aase, Methodological strategies in resilient health care studies: an integrative review. Safety Sci **110**, 300–312 (2018)
6. H.-F. Hsieh, S.E. Shannon, Three approaches to qualitative content analysis. Qual. Health Res. **15**(9), 1277–1288 (2005)
7. P. Mayring, Qualitative content analysis: theoretical foundation, basic procedures and software solution (2014)
8. M. Ekstedt, S. Ödegård, Exploring gaps in cancer care using a systems safety perspective. Cogn. Technol. Work **17**(1), 5–13 (2015)
9. M.D. Patterson, R.L. Wears, Resilience and precarious success. Reliab. Eng. Sys. Saf. **141**, 45–53 (2015)
10. L. Cuvelier, P. Falzon, Coping with uncertainty: resilient decisions in anaesthesia, in *Resilience Engineering in Practice: A Guidebook*, ed. by E. Hollnagel, J. Pariès, D.D. Woods, J. Wreathall (Ashgate, Farnham, 2011)
11. C.P. Nemeth, M. Nunnally, M.F. O'Connor, M. Brandwijk, J. Kowalsky, R.I. Cook, Regularly irregular: how groups reconcile cross-cutting agendas and demand in healthcare. Cogn. Technol. Work **9**(3), 139–148 (2007)
12. J. Paries, Resilience in intensive care units: the HUG case in resilient health care, in *Resilient Health Care*, ed. by E. Hollnagel, J. Braithwaite, R.L. Wears (Ashgate, Farnham, 2013)
13. A.S. Nyssen, A. Blavier, Investigating expertise, flexibility, and resilience in socio-technical environments: a case study in robotic surgery, in *Resilient Health Care*, ed. by E. Hollnagel, J. Braithwaite, R.L. Wears (Ashgate, Farnham, 2013)
14. B. Brattheim, A. Faxvaag, A. Seim, Process support for risk mitigation: a case study of variability and resilience in vascular surgery. BMJ Qual. Saf. **20**(8), 672–679 (2011)
15. V.J. O'Keeffe, M.R. Tuckey, A. Naweed, Whose safety? Flexible risk assessment boundaries balance nurse safety with patient care. Saf. Sci. **76**, 111–120 (2015)
16. M. Sujan, P. Spurgeon, M. Cooke, The role of dynamic trade-offs in creating safety - a qualitative study of handover across care boundaries in emergency care. Reliab. Eng. Sys. Saf. **141**, 54–62 (2015)
17. A.J. Ross, J.E. Anderson, N. Kodate, K. Thompson, A. Cox, R. Malik, Inpatient diabetes care: complexity, resilience and quality of care. Cogn. Technol. Work **16**(1), 91–102 (2014)
18. K.E. Weick, Collapse of sensemaking in organizations: the Mann Gulch disaster. Adm. Sci. Quaterly **38**(4), 628–652 (1993)
19. S.W. Dekker, J. Bergström, I. Amer-Wåhlin, P. Cilliers, Complicated, complex, and compliant: best practice in obstetrics. Cogn. Technol. Work **15**(2), 189–195 (2013)
20. E.S. Patterson, D.D. Woods, R.I. Cook, M.L. Render, Collaborative cross-checking to enhance resilience. Cogn. Technol. Work **9**(3), 155–162 (2007)
21. M.W. Smith, T.D. Giardina, D.R. Murphy, A. Laxmisan, H. Singh, Resilient actions in the diagnostic process and system performance. BMJ Qual. Saf. **22**(12), 1006–1013 (2013)
22. A. Miller, Y. Xiao, Multi-level strategies to achieve resilience for an organisation operating at capacity: a case study at a trauma centre. Cogn. Technol. Work **9**(2), 51–66 (2007)
23. K. Laugaland, K. Aase, The demands imposed by a health care reform on clinical work in transitional care of the elderly: a multi-faceted Janus, in *Resilient Health Care Volume 2: The Resilience of Everyday Clinical Work*, ed. by R.L. Wears, E. Hollnagel, J. Braithwaite (Ashgate, Farnham, 2015)

Chapter 11
Resilience from the United Nations Standpoint: The Challenges of "Vagueness"

Leah R. Kimber

Abstract A United Nations program, at the crossroad between the development and the humanitarian mandate (UNISDR) turned the concept of resilience into a central vehicle for its worldwide program on disaster risk reduction. It is through an ethnographic study of the negotiation process, topped by interviews and text analyses that I suggest various characteristics to describe resilience in an international organization. With the perspective of the sociology of translation, I discuss, on the one hand, the UN's need to maintain a vague definition of the concept, which hinders operationalization and on the other, I show how the organization manages, with resilience, to legitimize its programs and sustainability.

Keywords United Nations · International organizations · Resilience Terminology · Translation · Disaster · Risk · Reduction

11.1 Context and Introduction

This chapter examines the use of *resilience* at the United Nations (UN), in particular a UN program–United Nations International Strategy for Disaster Reduction (UNISDR)–which has the mandate to serve as "the focal point in the United Nations system for the coordination of disaster reduction and to ensure synergies among the disaster reduction activities of the United Nations system and regional organizations".[1]

In a nutshell, when it comes to studying *resilience* in the context of international organizations, it is associated with disasters. Building and increasing *resilience* is in part a question of reducing a population toward risks. In fact, the link between risk and resilience goes back to the paradigm shift in the 1980s as a result of the massive

---

[1] http://www.unisdr.org/who-we-are/mandate, consulted on 12 December 2017.

L. R. Kimber (✉)
Université de Genève, Geneva, Switzerland
e-mail: leah.kimber@unige.ch

© The Author(s) 2019
S. Wiig and B. Fahlbruch (eds.), *Exploring Resilience*, SpringerBriefs in Safety Management, https://doi.org/10.1007/978-3-030-03189-3_11

technological crises in the late 70s and early 80s. The notion of risk became integrated on the aftermath of these events across fields. Ulrich Beck [1] coined the term of the "risk society" era to account for this change. In that vein, many international actors got involved in managing the risks and the effects of natural hazards [2].

The growing incidences of death and destruction due to natural hazards have since played a major role for international organizations. The United Nations more specifically has managed to use the disaster narrative to legitimize programs and action plans, which promulgate norms and knowledge [2] and went as far as initiating the decade on international prevention for disaster reduction in the 1990s. The use of concepts such as vulnerability, disaster mitigation, disaster risk reduction and *resilience* within international organizations thus bear as witnesses for the paradigm change and the UN's concern to integrate the notion of "living with risk" in its programs.

While *resilience* was used in the context of hazards, crises and disasters in the late 1990s, it also became the outcome of *vulnerability*. Prior to *resilience*, *vulnerability* was key to studying natural hazards and poverty until the late 1980s, but was usually portrayed in negative terms as the susceptibility to be harmed [3]. *Resilience* rose as the positive replacement for vulnerability, which could be worked on and improved. UNISDR became the first UN agency to take on the *resilience* term, making it a central concept in its programs.[2]

From an organizational point of view UNISDR seeks legitimization to sustain itself. Institutionally, it does so by fulfilling a mandate at the crossroad between development and humanitarian international programs. Historically, the fracture between humanitarian organizations, dealing with emergency response, and development organizations, dealing with prevention, divides practitioners both at headquarters and in the field. "Humanitarian and development organizations tend to compete with one another for money, turf and credit" [4]. Nevertheless UNISDR focuses on decreasing vulnerability in the face of disasters from a development point of view and from a humanitarian perspective by linking climate change with Early Warning Systems (EWS).[3]

If we assume *resilience* has great potential, we question on the one hand whether it is operationalizable, and on the other, what it contributes to more broadly in international organizations. To do so, we follow the central concept at the heart of international negotiations from the perspective of the sociology of translation. It allows to pay close attention to the actors [5], which entails observing as far as possible, what they do as much as what they say [6] and gives an account of their arguments, points of views and contradictions. In other words, we translate the scallops' domestication and the fishermen of St Brieuc Bay— story explained in a groundbreaking article by

---

[2]Hyogo Framework for Action, 2005–2015 and Sendai Framework, 2015–2030.

[3]EWS: an integrated system of hazard monitoring, forecasting and prediction, disaster risk assessment, communication and preparedness activities systems and processes that enables individuals, communities, governments, businesses and others to take timely action to reduce disaster risks in advance of hazardous events to reduce losses. It also sets guidelines on how to best increase *resilience* from a preparedness standpoint (i.e. before a disaster strikes). It is thus viewed as an "a priori" outlook onto disaster.

[7]—to the one posed by the concept of resilience in the lead-up to UNISDR's text by Callon ratification on disaster risk reduction. The St Brieuc Bay actors encompassed the three main scientists, the fishermen, the scientific peers and the scallops all at play to discuss the ways in which they can improve scallop productivity. In the case of UN negotiations we find "our" main actors to be Member state representatives, UNISDR's staff member and members of Civil society. The "translation" thus needs to be viewed as a process in which a network of human and non-human actants (i.e. scallops being the *resilience* concept) construct common meanings and negotiate to reach individual or collective objectives [8]. We therefore look into UNISDR's *interessement* [7] in its attempt to impose and stabilize the other actors (Civil society and Member states) around defining *resilience*. Applying the sociology of translation to international organizations goes in line with Law's definition of organizations: precarious entities that require permanent stabilizing and ordering to maintain their actorness [9, 10].

11.2 Methodology

In order to analyze the purpose and operationalization limits of *resilience*, I rely on ethnographic data as well as content analysis [11]. *Embedded* [12–14] as member of the Women's Civil society group in the run up to the Third World Conference on Disaster Risk Reduction (WCDRR) held in Sendai, Japan in March 2015, I took notes and carried out 40 interviews during the negotiation process on the Sendai Framework text. Two ratified texts (Hyogo Framework for Action 2005, and Sendai Framework 2015) as well as interview and observation transcripts, as a result of a multi-sited ethnography [15] are at the heart of content analysis. I carried out a systematic search of the *resilience* concept to analyze what actors said about it and what the texts reveals. This in turn, allows me to look into the various uses of *resilience* throughout a UN agency, by providing a number of characteristics.

11.3 What Resilience Does at the United Nations

11.3.1 Creating "Vagueness"

In this section, I suggest to go over the various characteristics of *resilience* depicted at the UN to highlight the relevance of such a concept in international organizations. According to the analyses, five characteristics support the idea of *resilience* being a vague concept. I describe the forms *resilience* takes and illustrate the way the UN staff, Member states and Civil society members interpret and translate the concept.

First, I propose to associate *resilience* with the term *boundary object* [16]. In line with Brand and Jax's argument [17], who suggest the use of *resilience* as a facilitator within the field of science and technology, *resilience* facilitates communication across disciplinary borders. However, while easing communication, *boundary objects* also allow divergent meanings among the parties without it being necessarily openly recognized. Transposed to the UN context, *resilience* is a *boundary object*, which ties to tie two main disparities. On the one hand, it bridges the humanitarian and development divide among agencies enabling UNISDR to link both mandates [18] because both use it. On the other, the concept gives governments, UNISDR and Civil society the possibility to agree on common ground during negotiations.

> [Resilience] is one of these empty concepts really. It's whatever you want it to make it. [...] It's just a word. You can define and apply it in different ways. [...] Same as sustainability. Those big words they are kind of empty vessels and you put in them what [you want]. (Katherine, Women's Group member, 8.2.2016).

Second *resilience* participates in the constructive ambiguity *game* that is often times played among governments. This concept is similar to the abovementioned *boundary object*, or that of *flexible language* [19], but draws its roots from a different literature, that of international relations. In other words, ambiguity leads to greater leeway in implementation, because states end up circumventing obligations under other agreements and improve their negotiating positions in other ongoing processes [20], in [21]. *Resilience* is thus made ambiguous.

> We not all have the same recipe for resilience. Each country has its own capacity. There's not one formula. It's such a broad concept. (USA representative, 13.12.2016).

Third, *resilience* can be seen as a snake biting its tail. By analyzing the Hyogo Framework for Action text (HFA), *resilience* appears on many occasions and is understood as equivalent to "building a culture of safety" or even "a culture of disaster prevention". According to UNISDR, addressing "disaster risks" allows in turn "to manage and to reduce" them. Governments are thus encouraged to instill a set of means to stimulate a "culture of disaster resilience" and attain these by "developing and strengthening institutions", "enhancing governance for disaster risk reduction". Member states also need to use "innovation", "education and knowledge" more specifically knowledge pertaining to "hazards and the physical, social, economics and environmental vulnerabilities to disasters" by "promoting the engagement of media and food security" and ensuring that all "new hospitals are built with a level of resilience". Implementing these means fills the promise of a resilient outcome. Yet *resilience* also seems to be a means in itself. It is stated "disasters can be substantially reduced if people are informed and motivated towards a culture of disaster prevention and resilience". For UNISDR, the goal (reduce disasters) is reached providing there is *resilience*. While simultaneously acting as a mean, a goal and an outcome, *resilience* is hardly dissociable.

Fourth, *resilience* never appears as a stand-alone concept. It is associated with various words and tied to major concepts used at the UN in contexts of natural disasters. In the HFA, for example, we encounter "disaster resilience", "build a culture of safety and resilience", "culture of disaster prevention and resilience", "building resilience". In the Sendai Framework (SF), we come across "educational resilience of persons", "to promote a culture of disaster prevention, resilience and responsible citizenship", "economic, social, health and environmental resilience", "disaster risk resilience", "ensure resilience to shocks" and many more. In a context where *resilience* can at any time be juxtaposed to other concepts, it threatens the very definition and thus endorses vagueness.

Fifth, *resilience* contains within itself an irrevocable paradox; on the one hand *resilience* can only be attained by being creative [22] and on the other, UNISDR provides a framework. By framing and giving guidelines, we loose the fundamental component inbuilt in the *resilience* definition, namely that of creativity. In this sense again, a paradoxical statement leads to misconception and vagueness.

While UNISDR works toward reducing life-loss and limiting destructive outcomes, *resilience* does not appear as an operationalizable concept. The lack of a clear definition, which makes the concept persistently vague, hinders disaster risk reduction operationalization at headquarters as well as in the field. This having been said, even though these characteristics may be perceived as taking a toll on the worldwide programs, its vagueness seems to have a purpose. Further outlooks give cues on how *resilience* favors the organization's legitimacy.

11.3.2 Resilience as a Legitimizing Tool

While *resilience's* vagueness appears when confronting views on meaning and definition, other characteristics come to light. Here, I analyze the characteristics that convey a concept as a vector for an organization's sustainability.

Firstly, by acting as a non-controversial concept, *resilience* fosters international consensus within negotiation rooms as delegates express their desire to limit natural disaster impacts. The topic does not trigger salient political debate. On the contrary, in negotiation sessions with Member states, *resilience* is hardly discussed. Its definition[4] is a result of UNISDR's suggestion and not further discussed.

> [Resilience] is a word, I would say, yes, we all want to be resilient. It's like we all want to be sustainable (Marie, UNISDR staff, 9.02.2016).

Secondly, over the years, *resilience* turned into the positive outlook of *vulnerability*. If vulnerability gives the impression of a defined and static state — difficult to grow out of — *resilience* hints communities can work toward becoming stronger and more robust. *Resilience* thus turns into a driving force and a goal to reach, by providing a window of opportunity for change.

> [Resilience] is good in many ways because it allows us to focus less on vulnerability particularly in women and other gender groups and more on the positive and the capacities and capabilities etc. So it has value for me in that way because it should emphasize the positive

---

[4]"The ability of a system, community or society exposed to hazards to resist, absorb, accommodate to and recover from the effects of a hazard in a timely and efficient manner, including through the preservation and restoration of its essential basic structures and functions" (SF, 2015).

and how do you reach that positive stage rather than always focusing on the negative and the poor women. (Ellen, Women's Group member, 8.02.2016).

Thirdly, *resilience* comes across as an up-to-date word. *Resilience*, as a relatively new concept, emanates from other trendy concepts used earlier in the UN context. Indeed by taking on *resilience* and making it a central concept in a worldwide program, UNISDR jumps on the bandwagon in order to stay tuned with current jargon.

> Resilience is a good word just like sustainability is a good word. It's just like, in these international contexts, you need to change the terminology to keep it current, but really is reflecting many of the same things. So I still say disaster mitigation, but that's part of disaster risk reduction and that's part of... you know some people... I would write some documents where I would use some other terms, [such as disaster mitigation and I would be told] NO, [...] it is Disaster Risk Reduction. I was like "excuse me, I've been around!" (Cassandra, Women's Group member, 11.02.2016).

Finally, if we take a closer look into the late 2014, early 2015 negotiations in the run-up to the World Conference, *resilience* was hijacked by other international considerations. Even though Civil society strived to talk about disaster resilience and its practical considerations, Member states differed from the objective and raised politically charged issues. *Resilience* became of peripheral importance. It is the issues around "common but differentiated responsibility" (CBDR), "people living under the occupation", "technology transfer" that States inevitably raised causing lengthy negotiations. Nevertheless, it did not stop UNISDR from delivering a framework due to cover a time span of 15 years (2015–2030).

11.4 Conclusion

To conclude, integrating the *resilience* concept at the UN is not a naive undertaking. Rather it portrays an ideology, a vision of the world by using a certain language [23] namely that of making a world a better and safer place. While the initial motive seems to contribute to the reduction of disaster risk, it seems to overall serve a UN program. In particular, it helps in maintaining the organization's role and relevance regardless of the lack of operationalization. In this way, I explored the characteristics of *resilience* in international organizations, with a specific focus on the United Nations International Strategy for Disaster Reduction.

First, I presented five characteristics at the heart of the limits to operationalize *resilience*. Some of them stem from concepts in the literature (i.e. *boundary object, constructive ambiguity, flexible language*), others I developed for the purpose of this chapter (i.e. *a snake biting its tail, associated concept, non-controversial concept*). I showed how a vague definition creates challenges for future operationalization processes. Second, I put forward the characteristics, which play in favor of the organization's sustainability. Introducing a non-controversial concept allows to gather States, Civil society and various UN organizations to focus on how to positively impact consequences of disasters.

In this sense, studying the concept of *resilience*, in light of the actors at play, in a UN agency, accounts for the way issues are raised, the jargon updated and vagueness indispensable to keep the institution running. Even though *resilience* might come across as being a somewhat useless "empty vessel", it not only federates various bodies around one topic, but remains fundamental in that it allows a "sense of direction" in international negotiations.

> This is my view, but it's important to have a concept in order to have a direction even if it's blurry. Measuring it or at least trying to do so is a good sign. It shows that we are trying to go in that direction. I am a supporter of it. (George, UNISDR staff, 18.10.2016)

With this in mind, the chapter argues that such a concept participates not only in addressing the complexity of disasters and risk, but also plays a role in legitimizing the organization's role even in light of a lack of operationalizable targets. *Resilience* thus participates in legitimizing UNISDR's role as the main UN program for disaster coordination.

References

1. U. Beck, *Risikogesellschaft — Auf dem Weg in eine andere Moderne* (Suhrkamp Verlag, Frankfurt am Main, 1986)
2. S. Revet, « Vivre dans un monde plus sûr ». Catastrophes « naturelles » et sécurité « globale ». Cult. Confl. **75**, 33–51 (2009)
3. M.A. Janssen, E. Ostrom, Resilience, vulnerability, and adaptation: a cross-cutting theme of the international human dimensions programme on global environmental change. Glob. Environ. Change. **16**(3), 237–239 (2006)
4. J. Moore, The humanitarian-development gap. Int. Rev. Red Cross **81**(833), 103–107 (1999)
5. B. Latour, *Science in Action: How to Follow Scientists and Engineers Through Society* (Harvard University Press, Cambridge, 1987)
6. J. Best, W. Walters, Translating the sociology of translation. Int. Polit. Sociol. **7**(3), 345–349 (2013)
7. M. Callon, Some elements of a sociology of translation: domestication of the scallops and the fishermen of St Brieuc bay. Sociol. Rev. **32**(1), 196–233 (1984)
8. M. Wolf, A. Fukari (eds.), *Constructing a Sociology of Translation* (John Benjamins Publishing, Amsterdam, 2007)
9. J. Law, *Organising Modernity: Social Ordering and Social Theory* (Blackwell, New Jercy, 1994)
10. M. Müller, Assemblages and actor-networks: rethinking socio-material power, politics and space. Geogr. Compass **9**(1), 27–41 (2015)
11. P.Y. Martin, B.A. Turner, Grounded theory and organizational research. J. Appl. Behav. Sci. **22**(2), 141–157 (1986)
12. M. Bourrier, Pour une sociologie « embarquée » des univers à risque? *Revue de la société suisse d'Ethnologie* **15**, 28–37 (2010)
13. M. Bourrier, Embarquements. Socio-anthropologie **27**, 21–34 (2013)
14. M. Bourrier, Conditions d'accès et production de connaissances organisationnelles. Revue d'anthropologie des connaissances **11**(4), 521–547 (2017)
15. G.E. Marcus, Ethnography in/of the world system: the emergence of multi-sited ethnography. Annu. Rev. Anthropol. **24**(1), 95–117 (1995)

16. S.L. Star, J.R. Griesemer, Institutional ecology, 'translations' and boundary objects: amateurs and professionals in Berkeley's Museum of vertebrate zoology, 1907–1939. Soc. Stud. Sci. **19**(3), 387–420 (1989)
17. F.S. Brand, K. Jax, Focusing the meaning(s) of resilience: resilience as a descriptive concept and a boundary object. Ecol. Soc. **12**(1) (2007)
18. S. Revet, Les organisations internationales et la gestion des risques et des catastrophes naturels. Technical Report (Sciences Po, 2009)
19. K. Linos, T. Pegram, The language of compromise in international agreements. Int. Organ. **70**(3), 587–621 (2016)
20. O.R. Young, *The Institutional Dimensions of Environmental Change: Fit, Interplay, and Scale* (MIT Press, Cambridge, 2002)
21. I. Fischhendler, When ambiguity in treaty design becomes destructive: a study of transboundary water. Glob. Environ. Polit. **8**(1), 111–136 (2008)
22. K.E. Weick, Collapse of sensemaking in organizations: the Mann Gulch disaster. Adm. Sci. Quaterly **38**(4), 628–652 (1993)
23. T. Neeley, *The Language of Global Success: How a Common Tongue Transforms Multinational Organizations* (Princeton University Press, Princeton, 2017)

Chapter 12
Building Resilience in Humanitarian Hospital Programs During Protracted Conflicts: Opportunities and Limitations

Ingrid Tjoflåt and Britt Sætre Hansen

Abstract Humanitarian hospital programs supporting health systems during protracted conflicts require a combination of short- and long- term approach. Working in partnership, sharing of knowledge, provision of drugs, equipment and human resources together with a multi-sector and multilevel approach could contribute to build resilience in humanitarian hospital programs during protracted conflicts. However, withdrawal of humanitarian support after two years could lead to a possible decline in the quality of care linked to the end of the delivery of drugs, equipment and human resources if the local and national health authorities are not able to find any solutions to the chronic vulnerability. Continuous conflicts may continue to cause new challenges in these hospitals.

Keywords Humanitarian · Hospital programs · Protracted conflicts

12.1 Introduction

Protracted conflicts or recurring long-lasting conflicts based on one main conflict or from many different chronic conflicts gradually destroy infrastructure, services and living conditions. Not only a cause of human suffering, they are reasons for long-term displacement, migration and development delays. The character of these conflicts generates extreme fragility in basic services as well as in social, economic and environmental systems. The health systems in these conflicts are often overwhelmed as health professionals flee, infrastructure is destroyed and the provision of drugs and medical supplies is stopped [1]. Under these conditions, there are long-term humanitarian needs in terms of law and order, water, electricity, food security, health care and education. Humanitarian aid that supports health care in these conflicts requires a combination of short- and long-term approaches.

I. Tjoflåt (✉) · B. S. Hansen
Faculty of Health Sciences, University of Stavanger, Stavanger, Norway
e-mail: ingrid.tjoflat@uis.no

© The Author(s) 2019
S. Wiig and B. Fahlbruch (eds.), *Exploring Resilience*, SpringerBriefs
in Safety Management, https://doi.org/10.1007/978-3-030-03189-3_12

Building resilience has been raised as a new organizational principle by the United Nations, donors and Non-Governmental Organizations in development, climate change adaption and humanitarian aid [2, 3]. The key in building resilience is an attempt to reduce the dramatic decline in development and prevent unacceptable levels of human suffering that crisis and conflicts can cause [2, 4, 5]. However, it has been reported that it is not clear how resilience should be promoted during and after conflicts [6]. Frankenberger et al. [7] state that building resilience may be impossible where governments are fragile and where there are ongoing conflicts. The basic minimum conditions have to be present first. Nevertheless, it has been described that the humanitarian approach has to work with two perspectives simultaneously, that is, responding to immediate needs and mitigating the cumulative impact. The longer the conflict lasts, the more necessary it becomes to engage with people and communities at a structural level. Working in partnership is therefore an essential attribute in building resilience in humanitarian programs. Partnership reflects a participatory attitude built upon sensitivity, shared understanding and local knowledge [3, 8, 9]. There is always capacity in people or communities: to strengthen resilience is to increase this capacity. Coping and adapting are central aspects of resilience, and they refer to different actions that people do to mitigate difficulties or suffering. When resilience is understood as a capacity, it means that it is not a fixed concept but a dynamic one that varies and can change continuously. Consequently, it is then more relevant for humanitarian programs during protracted conflicts to focus on "building resilience" as a process that remains dynamic rather than trying to define and measure "resilience" as an outcome of an intervention [10].

There are numerous case examples from disaster areas and other situations of conflicts that describe the development and strengthening of national and local resilience capacities [11, 12]. As far as can be seen from the literature, there are no descriptions addressing the resilience in humanitarian hospital programs that improve the quality of care in local hospitals during protracted conflicts.

Therefore, this chapter is intended to shed light on the opportunities and the limitations of building resilience to improve the quality of care in local hospitals by implementing humanitarian hospital programs during protracted conflicts. Common areas of challenges in local hospitals during protracted conflicts will be described. Building resilience in the local hospital will then be discussed in relation to partnership, sharing of knowledge and provision of drugs, equipment and human resources.

12.2 Experiences from the Field

The first author has extensive experience working to improve the quality of care in local hospitals with a humanitarian organization in different protracted conflicts in Africa, Asia and the Middle East. Moreover, she has conducted qualitative research related to challenges in improving the quality of care in different humanitarian programs [13–15]. Based on the first author's experience, challenges local health personnel face working in hospitals during protracted conflicts are described. To emphasize the challenges and highlight important elements, the listed challenges have been simplified. Of note, the challenges does not refer to any specific hospital, but has been constructed from a range of experiences.

Common areas of challenges in local hospitals during protracted conflicts

- There are very few beds compared to the population.
- The wards are normally overcrowded and the mortality rate is high. Due to ongoing conflicts, patients often cause extra caseloads in the wards.
- There is an insufficient supply of drugs and disposable equipment to treat the patients. The patients and their relatives have to purchase drugs and equipment in the local pharmacy, which often delays treatment and results in poor quality of care. To reduce the patients' financial costs, the staff have to use as little equipment as possible when performing procedures.
- There is a lack of basic medical equipment, such as a blood pressure cuff, stethoscopes, oxygen and mattresses. Some of the equipment is broken. Even if some health personnel wanted to try to maintain the national standard of care, the lack of drugs, disposable equipment and medical equipment makes it impossible.
- Sometimes the hospital has problem with insufficient electricity and water.
- Maintenance of the hospital buildings has not been carried out for years.
- There is a lack of skilled staff because a majority has fled the country. In addition, due to the ongoing conflict there is not enough health staff educated.

Support by an international humanitarian organization

Sometimes humanitarian organizations provide support to one ward, for example to a surgical ward due to the extra pressure war and violence put on surgical services. Such support might include teaching/supervision and logistical support related to essential medical drugs and equipment with the aim of building resilience by improving the quality of care. Some maintenance support of the ward may also be carried out. Such support might be offered as a two-year partnership project. The general objective in such projects, based on a joint assessment by the hospital leadership and the humanitarian organization, might be that surgical patients will be cared for in a functional, well-equipped and maintained department with skilled health personnel who meet the relevant national standards of care. The international humanitarian organizations may send visiting specialist health workers, who might do short mission in the hospital and conduct training. The visiting health workers from the humanitarian organization may work alongside the local health staff to share knowledge to achieve a sustainable standard of treatment and care, comparable with the national standards in the country. In addition, the humanitarian organization may provide a program manager and one administrator to work with the management in the hospital for the length of the project.

12.3   Discussion

Based on the challenges listed above, three essential areas for building resilience in the local hospital surgical ward will be discussed. These are partnership, knowledge sharing and the provision of drugs, equipment and human resources. The discussion will focus on both the opportunities and the limitations in building resilience.

12.3.1   Partnership

One opportunity for building resilience is that the project is a partnership project. Not only is working in a partnership an essential attribute in building resilience in humanitarian programs, but it also reflects a participatory attitude [3, 8, 9]. The general objective of humanitarian projects are based on a joint assessment by the hospital leadership and the humanitarian organization. Jointly assessing and defining the general objective of a project together can provide opportunities for building resilience through a dialogue encompassing mutual respect and a willingness to listen and understand the local hospital's point of view [3, 8, 9].

However, a partnership project also requires a proper understanding of how the hospital system works and a deeper engagement with the structures in the hospital as well as with the health system in the country [1, 16]. Therefore, to build resilience in one ward or in a small part of a hospital, the humanitarian organization has to conduct a robust analysis that goes beyond needs assessments [12]. Moreover, the visiting health workers, the program manager and administrator from the humanitarian organization should listen to the local health workers and understand their perspectives as well as those of the hospital and the health system. One limitation in building resilience in such partnership projects is that the humanitarian support only focus on a small part of a hospital.

Forming a successful partnership takes time and expertise. Whether the humanitarian organization has the time and the knowledge to build a partnership is a question that must be asked. Based on the first author's experiences, it can be challenging to create a good partnership in areas of conflict due to security constraints, the immediate need to save lives, the limited time available to set up a program and/or a fragile and/or absent local management and authorities. These experiences are also reported in studies. A study that interviewed more than 6,000 people from around the world who have received international assistance showed that, all too often, trust and respect between partners can diminish during emergencies. The study reported that the local partners perceived the international agencies to be "paternalistic" in taking over local initiatives. Additionally, a lack of respect for and appreciation of local knowledge and contributions limited the extent of the partnerships, since the locals are rarely involved in the decision-making processes with their partners [17].

12.3.2   Knowledge Sharing

Another opportunity in the project for building resilience with regard to improving the quality of care is the sharing of knowledge and experience between the visiting health workers and the local staff in the surgical hospital ward.

To share knowledge, the visiting health workers must recognize and appreciate the efforts the health personnel working in the ward who have done what they could to maintain the quality of care during years of conflict. The described challenges reveals that, due to a lack of equipment, the staff had to find solutions to take care of patients compromising daily the national standard of care and treatment. The visiting health workers, who work alongside the local health staff in the surgical ward, should be able to identify and understand the local staff's efforts in trying to maintain standards of care and the visiting health workers should value the local knowledge and experience. The plans and efforts should be put at the service of the local health staff's initiatives and their capacities. This is emphasized in the literature and it serves to establish humanitarian policy documents as essential attributes for building resilience [3, 8, 9]. However, studies from humanitarian missions to conflict areas show that visiting health workers rarely acknowledge or use the local knowledge [15, 18, 19]. Tjoflåt et al. [15] reported that the visiting nurses admired the local nurses' creativity in work situations with few available resources. However, it was a challenge for the visiting nurses to utilize this knowledge when they worked together with local nurses in the ward to improve the quality of care. The visiting nurses had problems adhering to the local nursing standards, which they saw as substandard compared with their home country, and they took care of the patients themselves (ibid). The visiting nurses' negative attitude towards local standards and taking care of the patients themselves could possibly hamper the improvement of care in the surgical ward and further limit the building of resilience.

Without improving the availability of drugs and medical equipment and reversing the shortage of health personnel in the ward, knowledge sharing could have only a minimal impact on the quality of care for surgical patients.

12.3.3   Provision of Drugs, Equipment and Human Resources

Other opportunities for building resilience in the hospital's surgical ward come from the provision of essential drugs and equipment to the surgical ward for the length of the project. Moreover, the maintenance to the ward. Such support will certainly improve the quality of care and will provide opportunities for building resilience, but it does not solve the chronic vulnerability related to supply, which often is linked to ongoing conflicts.

Another limitation for building resilience may be the lack of human resources. The lack of qualified staff may be one reason for high mortality rate in these hospitals. For the length of the project, the visiting health workers will provide some extra health staff to the ward when they work alongside the local staff to share knowledge, but the quality of care may decline when the project ends if the national health authorities fail to increase the number of staff permanently.

Building resilience in these areas requires a multi-sector and multilevel approach. Different types of interventions and sequentially have to be addressed with various department in the hospital as well as with the management of the hospital to enable resilience related to provision of drugs, equipment and human resources. The lack of supply and human resources in the hospital can be reported to the national health authorities by the management of the hospital with the support of the international humanitarian organization, but the national authorities have to find long-term solutions to ensure sufficient supplies and human resources. The *European Commission's Action Plan for Resilience in Crisis Prone Countries 2013–2020* reports that building resilience is a long-term process and should be embedded in national policies and planning. Where the state and situation are fragile, it is essential to identify functioning systems within local institutions and support their capacity.

Another limitation in the humanitarian support for building resilience is the length of the project. Closure of the partnership project after two years and ending the provision of essential drugs, equipment and the human resource support to the surgical ward may result in a decline in the quality of care if no other solutions are found for example an extension of the project. To build resilience often requires a long-term engagement and investment [4].

12.4   Conclusion

The discussion of the experienced challenges revealed that partnership, knowledge sharing, the provision of drugs, equipment and human resources together with a multi-sector and multilevel approach could contribute to build resilience in humanitarian hospital programs during protracted conflicts. However, withdrawal of humanitarian support after two years could lead to a possible decline in the quality of care linked to the end of the delivery of drugs, equipment and human resources if the local and national health authorities are not able to find any solutions to the chronic vulnerability. On the other hand, limiting the support to single wards or smaller part of local hospitals during protracted conflicts may partial support the hospitals' resilience. Continuous conflicts may continue to cause new challenges in these hospitals. Realistically it will not be possible for a humanitarian organization to support all challenges in local hospitals during protracted conflicts. Priorities must be determined according to the context and resources available.

References

1. ICRC, Protracted conflict and humanitarian action: some recent ICRC experiences. Technical report (International Committee of the Red Cross, 2016)
2. Action plan for resilience in crisis prone countries 2013–2020, Commission staff working document, European Commission (2013)
3. UN, The Sendai framework for disaster risk reduction 2015–2030. Technical report (United Nations, 2015)
4. IFRC, The road to resilience: bridging relief and development for a more sustainable future. Technical Report (International Federation of Red Cross and Red Crescent Societies, 2012). IFRC discussion paper on resilience
5. S. Levine, I. Mosel, Supporting resilience in difficult places: a critical look at applying the "resilience" concept in countries where crises are the norm. Technical report (Humanitarian Policy group, Overseas Development Institute, 2014)
6. S. Levine, A. Pain, S. Bailey, L. Fan, The relevance of "resilience"? HPG Policy Brief, vol. 49 (Humanitarian Policy Group, 2012)
7. T. Frankenberger, T. Spangler, S. Nelson, M. Langworthy, Enhancing resilience to food insecurity amid protracted crisis (2012)
8. C. Byrne (ed.), Participation crisis-affected populations in humanitarian action: a handbook for practitioners overseas (Overseas Development Institute, 2003). The Active Learning Network for Accountability and Performance in Humanitarian Action (ALNAP)
9. GPH, The global humanitarian platform's principles of partnership (2007)
10. S.B. Manyena, The concept of resilience revisited. Disasters **30**(4), 434–450 (2006)
11. EU resilience compendium saving lives and livelihood. Technical report, European Commission Humanitarian Aid and Civil, Protection DG (2015)
12. IFRC, World disaster report (2016) Resilience: saving lives today, investing for tomorrow. Technical report (International Federation of Red Cross and Red Crescent Societies, 2016)
13. I. Tjoflåt, B. Karlsen, Challenges in sharing knowledge: reflections from the perspective of an expatriate nurse working in a South Sudanese hospital. Int. Nurs. Rev. **59**(4), 489–493 (2012)
14. I. Tjoflåt, B. Karlsen, Building clinical practice in the Palestine Red Crescent operation theaters in Lebanon: reflections from the perspective of an expatriate nurse. Int. Nurs. Rev. **60**(4), 545–549 (2013)
15. I. Tjoflåt, B. Karlsen, S.B. Hansen, Working with local nurses in order to promote hospital-nursing care during humanitarian assignments overseas: experiences from the perspectives of nurses. Jn. Clin. Nurs. **25**, 1654–1662 (2016)
16. E. Hollnagel, Making health care resilient: from safety -1 to safety –ii, in *Resilient Health Care*, ed. by E. Hollnagel, J. Braithwaite, R.L. Wears (Ashgate, Farnham, 2013)
17. M.B. Anderson, D. Brown, I. Jean, Time to listen: hearing people on the receiving end of international aid. CDA Collaborative Learning Projects (2012)
18. M. Girgis, The capacity-building paradox: using friendship to build capacity in the South. Dev. Pract. **17**(3), 353–366 (2007)
19. A.L. Martiniuk, M. Manouchehrian, J.A. Negin, A.B. Zwi, Brain gains: a literature review of medical missions to low and middle — income countries. BMC Heal. Serv. Res. **12**(134) (2012)

Chapter 13
Exploring Resilience at Interconnected System Levels in Air Traffic Management

Rogier Woltjer

Abstract This chapter raises issues and ideas for exploring resilience, stemming from various research disciplines, projected on the domain of air traffic management and aviation at interconnected system levels. Attempts are made to connect micro, meso, and macro levels in the aviation sector identifying corresponding research challenges. Examples of this ongoing research are given on how theory has already been translated into practical methodological use. Some connections between foci from Resilience Engineering, Disaster Resilience, and other research disciplines are projected on the air traffic management domain to explore how practical benefits can be obtained from these theories and which aspects of operational practice these theories connect to. The chapter shows that the concept of resilience from various research disciplines has a potentially wide application to system levels of air traffic management, and suggests resilience to be addressed from an interconnected systems perspective to provide added value to operations.

Keywords Air traffic management · Resilience · Cross-scale interactions
Systemic · Interconnected · Systems

13.1 Introduction

This chapter seeks to explore the added value of the concept of some of the definitions that frame resilience as kinds of adaptive capacity, and addresses resilience-related aspects at interconnected micro, meso, and macro levels in Air Traffic Management (ATM). The chapter draws upon several research projects on method and guideline development in ATM that focuses on resilience as adaptive capacity and performance variability management as well as resilience as related to crisis management.

R. Woltjer (✉)
Swedish Defence Research Agency (FOI), Stockholm, Sweden
e-mail: rogier.woltjer@foi.se

© The Author(s) 2019
S. Wiig and B. Fahlbruch (eds.), *Exploring Resilience*, SpringerBriefs
in Safety Management, https://doi.org/10.1007/978-3-030-03189-3_13

The SESAR 16.1.2 project developed and applied principles from the Resilience Engineering literature to safety assessment and design of future technical and operational concepts for air traffic management [1]. The DARWIN project has conducted a worldwide systematic literature review covering more than 400 articles related to resilience and critical infrastructure [2]. It aims to develop resilience management guidelines [3] and adapt these to health care and air traffic management. The systematic literature review identified resilience research on micro, meso, and macro levels as well as on resilience in response to a variety of circumstances, from uncertainty and change, to disruptions and crises, to everyday variability. Related work on agile inter- and intra-organisational response to various crises in the aviation industry [4] may also be mentioned in this respect.

13.2   The Added Value of the Term Resilience

The literature is highly diverse in its use of the term resilience [2]. The discussion in this chapter is in line with the position of Woods [5] that resilience is conceptually different from the terms rebound and robustness, although the terms are oftentimes used synonymously to resilience. Among other reasons, these terms lack the adaptive capacity component of resilience, which in this chapter is seen as a salient aspect of the use of resilience in general and for Air Traffic Management specifically.

This chapter will first try to determine the "added value" of using the term resilience rather than other terms or other uses of the term, by briefly addressing a few of the definitions and descriptions of resilience that take adaptive capacity as a central theme. These stem from Resilience Engineering, mostly based on human factors and safety management, and Disaster Resilience originating in crisis and disaster management.

Resilience as "the intrinsic ability of a system to adjust its functioning prior to, during, or following changes and disturbances, so that it can sustain required operations under both expected and unexpected conditions." [6, p. xxxvi] implies standpoints of Resilience Engineering relative to traditional approaches to safety [1]. It emphasises the need to not only react and respond when disturbances are observed but also when they are *anticipated* to occur. Adjusting performance not only in relation to disturbances but also more subtle *changes* is highlighted, as common everyday fluctuations in working conditions may coincide and coalesce [6] to hazardous situations, due to system complexity. The terms 'required operations' and 'functioning' emphasise the need to appreciate the *multiple goals* that operations try to balance. This includes not only safety but often also productivity/profit, efficiency, security, environmental sustainability, etc. Referring to both the 'expected' and 'unexpected' emphasises the need to recognize that not all conditions can be expected and prepared for beforehand, and that *unexpected conditions* are likely to transpire in complex systems. Traditional approaches to safety focus on anticipation and mitigation of risks, i.e., preparing for the expected. Resilience Engineering suggests that working conditions and processes may be designed to support coping with unexpected events.

This approach was further pursued in the SESAR 16.1.2 project, which adapted the definition above to the following: "The intrinsic ability of the ATM/ANS functional system to adjust its functioning and performance goals, prior to, during, or following varying conditions" [1, p. 120]. Several key aspects led to this refinement. First, the language needed to be adjusted to fit the SESAR safety assessment methodology, a method based on more traditional safety engineering, which the project was tasked to feed Resilience Engineering methodology into.

Second, *performance goals change* depending upon the situation, which was a refinement of Hollnagel's "required operations". For example [1], everyday Air Navigation Service aims to provide both safe and efficient flow of traffic, but this, in case controllers' traffic display blacks out, moves to mainly providing separation (safety) between aircraft using all means available. Third, the term varying conditions is among other objectives aimed at including what in Hollnagel's definition is called expected and unexpected in one phrase. In many cases there are *both expected and unexpected aspects* to a complex air traffic situation or course of events, so varying conditions is less divisive or binary and thus captures various degrees of regularity of conditions.

This leads to another connection to traditional safety engineering, as the conditions that have not been documented in safety cases could be called unexpected conditions. These could however simply be events that indeed have been thought of during development work but did not need to be addressed further due to too low probability and/or severity in the risk matrix to make it to formal addressing as part of regulated safety assessment. Another issue is that "the operational situation" at any point in time is a complicated aggregate of many conditions, some of which are expected, important, challenging and/or meaningful, or not, depending on who you ask. Also, in complex domains including ATM [1] it seems to be evident that coping with everyday situations (which include unexpected conditions) is based on controllers' ability to apply and merge previous experience with preparations for expected conditions (design features, training, procedures, etc.) in new ways. In sum, the need for addressing expected and unexpected conditions simultaneously, in some way connecting to traditional ways of addressing expected conditions, while not oversimplifying complexity and diversity, but maintaining practicality of the approach with limited resources for assessing any future ATM concept, is thus a challenge when introducing resilience perspectives.

In order to achieve resilience, four necessary and interacting abilities have been defined as anticipating (knowing what to expect), monitoring (knowing what to look for), responding (knowing what to do), and learning (knowing what has happened) [7]. This thus ties together several established activities of traditional safety management, such as risk analysis and assessment, safety oversight, safety indicators, and incident investigation. These activities have traditionally been focused on failures and kept mostly separate from business management processes. Resilience Engineering has a different perspective on this focus, emphasizing that all processes and outcomes of everyday performance, productivity, and safety need to be understood from an integrative management perspective.

Articulating the importance of unexpected conditions in Resilience Engineering, Woods' 2006 definition focuses on the situations that go beyond what the organisation or system has prepared for: "the ability to recognize and adapt to handle unanticipated perturbations that call into question the model of competence, and demand a shift of processes, strategies and coordination" [8, p. 22]. This definition seems to include only unexpected beyond-design base processes and strategies in resilience. Although this seems like a welcome distinction in order to avoid that resilience is used as an overly inclusive term, the previous discussion of combinations of various degrees of expectation in a complex combination of varying conditions seems to imply that the distinction of when a competence model is called into question, and what constitutes a shift and what does not, could be difficult to make practically.

The emergency and disaster management literature has acknowledged the potential contribution of the concept of resilience for some time [9]. For example, Comfort, Sungu, Johnson, and Dunn [10] discuss public organisations in risky dynamic environments. They emphasise these organisations' need for a balance between anticipation, meaning assessment of vulnerability and safety and (planning for) preventive action, and resilience, meaning (planning for) flexible response ('bouncing back') after a damaging event [10]. Disaster resilience authors seem to resonate with Woods' suggestion to limit resilience to unanticipated conditions requiring adaptation [11].

Although years of theorizing have passed, agreement on the term resilience in the short term in the broad set of research communities using the term [1] seems unlikely. For new concepts like resilience to contribute beneficially to operations, these distinctions may however be important if the concept is to find its way into concrete methods in highly regulated domains such as ATM.

13.3   Micro Level Resilience: The Controller

At a micro level the resilience of the controller may be addressed from at least two perspectives: The psychological processes involved in the well-being of the controller in handling disturbances, and the cognitive processes involved in the actual controlling of the traffic. The latter is arguably most effectively addressed at the meso-level instead of the micro-level, where the controllers and the technical tools available to them in a joint human-human-tool-system can be described as performing cognitive tasks in terms of functional units. The former is currently mostly addressed as part of the work on Critical Incident Stress Management (CISM), which has found its way from other domains into ATM [12]. CISM is a peer support program that has as its objectives to mitigate the effects of harmful events, facilitate recovery, restore adaptive functioning, and identify who would benefit from additional services or treatment. As Mitchell & Leonhardt [12] describe, it is a multi-faceted flexible approach with a number of anticipatory, monitoring, and pedagogical activities before a harmful event occurs, support as a response immediately after a harmful event, as well as longer-term recovery. As such, it is an example of increasing the resilience at the micro-level through engagement at the meso-level, i.e. a set of practices between controllers and organizational processes covering all of Hollnagel's resilience cornerstones in order for the individual controller to be psychologically resilient. Thus it is an example of micro-meso interaction that can be argued necessary to establish micro-level (controller) resilience as well as entailing meso-level resilience in the sense that an ATM organization at the local level organizes peer support for all controllers. As such the mechanisms of adaptive capacity and adaptive performance may though not be easily observable or measurable because of the difficulty in assessing purely psychological processes, and need to be addressed *at micro and meso levels simultaneously*.

13.4 Meso Level Resilience: The Position, Sector, and Tower/Center

The meso level in ATM may be seen as containing several gradations in units of analysis, at least those at the controller working positions at the sector that is controlled, as well as at the level of the air traffic service units (ATSU) where simply put three different kinds of services are provided: tower control (in the control zone directly around the tower on an airport), terminal control (the wider area around the control zone used for flights approaching and departing airports), and area control (for controlling flights en-route or climbing/descending to and from terminal areas).

At the area controller working position, services are typically provided to aircraft by two controllers that work on a sector of airspace with a limited number of aircraft, with an "executive" controller who talks to the pilots and a "planner" controller that provides help to the executive controller by coordinating with other sectors, anticipating traffic and pre-emptively implementing or suggesting solutions, etc. Tower and terminal controllers may also work on several closely coordinated working positions. In this way services such as separation and route management, sequencing, and handling pilot requests are handled. To their aid the controllers typically have a suite of technical systems, besides the situation display showing the traffic also for the management of separation between aircraft, planning sequences of traffic, traffic conflict management, anticipation of routing consequences for efficiency and safety, etc. As the functions of separation management and planning for example are jointly performed by both executive and planner controllers with the continuous help of their technical tools, the "cognitive" functions of decision making, planning, attention management, etc., are most meaningfully addressed at the "joint cognitive system" [13] level of the sector. Resilience of the operational activity of handling the air traffic should thus likely also be addressed at this level. In terms of Hollnagel's resilience cornerstones, anticipation, monitoring, and providing a response (of/to traffic and its dynamic behaviour) is performed jointly by several controllers and their technical systems in a highly intertwined manner, i.e. most actions taken to handle air traffic originate from the sum of system parts.

Aggregating sectors then brings the perspective at the ATSU level, consisting of a number of sectors with the setup as described above complemented with operational and technical supervisors, with sectors dynamically grouped or split depending on changing traffic demand and circumstances such as weather. Some characteristics of the activities involved in anticipating, monitoring, and responding to events, and especially learning, cannot meaningfully be expressed at lower levels than the grouped sector or ATSU level. In case of an unexpected event happening in one sector, for example where large amounts of traffic need to be rerouted, help can often be obtained from other sectors, so that several sectors are involved in rerouting and helping each other in handling the traffic jointly. This is however not only reactively, in some cases this can go as far as controllers spotting potential problems for controllers numerous sectors ahead for traffic in their sector. Regular cooperation between adjacent sectors on the boundaries between ATSU areas of responsibility, such as between area control centers or countries, leads to the necessity to address adaptive capacity at the inter-ATSU (macro) level. This means that in these cases adaptive capacity cannot adequately be understood unless activities at several ATSUs (towers, terminal/area control centers, possibly in different countries) are addressed. Moreover, the resilience at the meso-level in air traffic management also needs to include other operators than air traffic controllers representing other stakeholders in the air traffic system, such as supervisors, technicians, pilots, ground vehicle operators, etc.

At the sector and ATSU levels, the SESAR 16.1.2 project [1] generated guidance on how to address resilience from a Resilience Engineering perspective (using eight principles mostly based on the work by Hollnagel and Woods): work-as-done, addressing the ways operators use procedures and other working methods, strategies and practices to achieve safety and efficiency, and to meet varying conditions, interpreting signals and cues, trying to find balance in goal trade-offs, while providing adaptive capacity, coping with complex couplings and interactions, managing timing, pacing, and synchronization, in an environment with under-specification and making necessary but approximate adjustments.

Thus the meso-level has several distinctions in grouping of relevant units of analysis which need to be addressed and understood in concert in order to understand ATM resilience and how the system provides adaptive capacity. Principles originating from the Resilience Engineering literature have been operationalized in the ATM domain. Most of these principles *apply to several meso-levels and should be addressed at these levels jointly* in order to obtain a comprehensive resilience perspective.

13.5 The Macro Level: National and International Organizations

At the macro level which here is identified as the national and international (societal) level, ATM is first of all an international network of ATSU nodes cooperating and collaborating where needed to handle air traffic. A number of organizations exist such as the Network Manager responsibility assigned to Eurocontrol which performs a number of operational functions to increase safety and efficiency of the aviation network as a whole. Addressing the resilience of the European ATM network or the European air traffic system would not be meaningful without addressing these aspects. This can also be said for handling large-scale crisis events: The European Aviation Crisis Coordination Cell (EACCC) coordinates the management of crisis response in the European ATM network. The main role of the EACCC is to support coordination of the response to network crisis situations, in close cooperation with the corresponding national crisis response functions and agencies, including coordinating responses and facilitating information sharing. Thus, when crises are of a scale (possibly after escalation within aviation, or from or to other industries or parts of society), *addressing resilience even at a meso level* (of sectors and ATSUs) *need to include an understanding at the macro level activities* (for example rerouting of traffic between countries and restrictions on traffic load affecting several countries). A particularly clear example of this are the events after the 2010 eruptions of the Islandic Eyjafjallajökull volcano which disrupted air travel across Europe, and affected several other means of transportation.

13.6   Discussion

This chapter has aimed to show that resilience in ATM, and arguably in other safety-critical network-based parts of industry and society, needs to be understood and addressed at micro, meso, and macro scales appreciating interconnectedness and cross-scale interactions [8]. Resilience also pertains to many functional levels and system groupings, making even a distinction like micro, meso, and macro difficult to specify distinctly, due to the networked nature of ATM. Due to this diversity and wide applicability of the approach, the definition of resilience becomes important, yet the identification of resilience characteristics and especially metrics and measurements is particularly challenging. By studying resilience at these diverse interconnected levels and establishing a vocabulary strongly connected to the operational vocabulary at different scales, resilience research may contribute to a better understanding of adaptive capacity and coping with our increasingly complex world.

Acknowledgements This work is among other sources based on the author's work in the SESAR 16.1.2, and DARWIN projects. Project partners are thankfully acknowledged. However, opinions in this chapter reflect only the author's views and are not intended to represent the positions of these projects' sponsors, member organizations or partners. The research leading to these results received partial funding from the European Union's Horizon 2020 research and innovation programme under grant agreement No 653289 (DARWIN). Any dissemination reflects the author's view only and the European Commission is not responsible for any use that may be made of the information it contains.

References

1. R. Woltjer, E. Pinska-Chauvin, T. Laursen, B. Josefsson, Towards understanding work-as-done in air traffic management safety assessment and design. Reliab. Eng. Sys. Saf. **141**, 115–130 (2015)
2. Consolidation of resilience concepts and practices for crisis management (DARWIN Deliverable D1.1). Technical report, DARWIN project (2015)
3. M. Branlat, R. Woltjer, L. Save, O. Cohen, I. Herrera, Supporting resilience management through useful guidelines, in *Proceedings of the 7th Resilience Engineering Association Symposium* (2017)
4. R. Woltjer, B.J.E. Johansson, P. Berggren, An overview of agility and resilience: from crisis management to aviation, in *Proceedings of the 6th Resilience Engineering Association Symposium* (2015)
5. D.D. Woods, Four concepts for resilience and the implications for the future of resilience engineering. Reliab. Eng. Sys. Saf. **141**, 5–9 (2015)
6. E. Hollnagel, Prologue: the scope of resilience engineering, in *Resilience Engineering in Practice: A Guidebook*, ed. by E. Hollnagel, J. Pariès, D.D. Woods, J. Wreathall (Ashgate, Farnham, 2011), pp. xxix–xxxix
7. E. Hollnagel, Epilogue: RAG - the resilience analysis grid, in *Resilience Engineering in Practice: A Guidebook*, ed. by E. Hollnagel, J. Pariès, D.D. Woods, J. Wreathall (Ashgate, Farnham, 2011)
8. D.D. Woods, Essential characteristics of resilience, in *Resilience Engineering: Concepts and Precepts*, ed. by E. Hollnagel, D.D. Woods, N. Leveson (Ashgate, Aldershot, UK, 2006), pp. 21–34
9. S.B. Manyena, The concept of resilience revisited. Disasters **30**(4), 434–450 (2006)
10. L.K. Comfort, Y. Sungu, D. Johnson, M. Dunn, Complex systems in crisis: anticipation and resilience in dynamic environments. J. Contingencies Crisis Manag. **9**(3), 144–158 (2001)
11. A. Boin, L.K. Comfort, C.C. Demchak, The rise of resilience, in *Designing Resilience: Preparing for Extreme Events*, ed. by L.K. Comfort, A. Boin, C.C. Demchak (University of Pittsburgh Press, Pittsburgh, 2010)
12. J.T. Mitchell, J. Leonhardt, Critical incident stress management (CISM): an effective peer support program for aviation industries. Int. J. Appl. Aviation Stud. **10**(1), 97–116 (2010)
13. E. Hollnagel, D.D. Woods, *Joint Cognitive Systems: Foundations of Cognitive Systems Engineering* (CRC Press, Boca Raton, 2005)

Chapter 14
Resilience in Healthcare: A Modified Stakeholder Analysis

Mary Chambers and Marianne Storm

Abstract *Resilient healthcare* embraces complexity, performance variability and acknowledgement of when things go right and when things go wrong it is usually because there has been an aspect of organizational malfunction or failure. Each organisation comprises of a range of stakeholders both internal and external and holding a variety of roles. To gain a better understanding of how individuals and groups influence the decision-making process of organisations, a *stakeholder analysis* can be the appropriate approach of choice. This chapter presents an approach to stakeholder analysis within the context of health care and the growing realization that patients and pubic can make a valuable contribution to the decision-making process of organisations and the contribution to *resilient health care*. Highlighted within the chapter are key questions and stages that require consideration when conducting a *stakeholder analysis*. To incorporate the contribution of patients and public, we use an analytical framework describing different aspects (decisions-making domains, roles and levels) of participation in healthcare decision-making. Reference is made to the benefits of conducting a *stakeholder analysis*, what the results can contribute with and indicates some of the challenges.

Keywords Resilience · Stakeholder analysis · Mental health · User involvement

14.1 Introduction

Resilient healthcare embraces complexity, performance variability and acknowledgement of when things go right [1]. Going right means that the system functions as it should and people work-as-imagined; when things go wrong it is because something has malfunctioned or failed.

M. Chambers (✉)
Kingston University & St. George's University of London, Cranmer Terrace,
London SW17 0RE, UK
e-mail: M.Chambers@sgul.kingston.ac.uk

M. Storm
University of Stavanger, Stavanger, Norway

© The Author(s) 2019
S. Wiig and B. Fahlbruch (eds.), *Exploring Resilience*, SpringerBriefs
in Safety Management, https://doi.org/10.1007/978-3-030-03189-3_14

One factor that can contribute to the development of *resilient healthcare* is service user involvement and engagement of patients and service users in decision making processes at the individual and organizational level of healthcare services. User involvement can, which can impact on how knowledge is shared and situations of adversity managed.

In this chapter, we will focus on a modified *stakeholder analysis* [2], of *resilient healthcare* taking mental health care and service user involvement as a case example. To guide the analysis we will draw on aspects of the analytical framework proposed by [3] as they describe different aspects of lay participation in healthcare decision-making.

Definition and Categorization of Stakeholders. A stakeholder can be defined as "persons or groups that have, or claim that they have, ownership, right or interests, in a cooperation and its activities, past, present and future" [4] cited in [2]. Stakeholders can be categorized in terms of how they interact with organizations, for example they can be internal, operate at the organizational interface or external [2]. Reference [4] refers to primary and secondary stakeholders and considers both as essential to the functioning of an organization. With respect to healthcare organizations, primary stakeholders can be viewed as patients, their families or next of kin, service users, clinical staff (e.g. nurses, medical doctors), administrative personnel and organizational leaders. The secondary stakeholders are those, who interact with the organization, but are not essential to the organizations existence such as voluntary groups or other support organisations.

14.2 Stakeholder Analysis

There is growing interest in the use of *stakeholder analysis*. This is reflective of the increasing awareness of how stakeholders, for example, groups and individuals can influence the decision-making process regarding the delivery, (in this instance) of healthcare and how individuals can inform and shape the nature of their own care and treatment.

A *stakeholder analysis* is an approach, a tool or set of tools for generating knowledge about individuals or organisations to better understand their behavior, intentions, inter-relationships and interests giving consideration to the influences and resources they bring to bear on decision making and/or implementation processes [2].

The purpose of a *stakeholder analysis* is to help understand stakeholders from the perspective of an organization or to determine the relevance of stakeholders to a particular research project, quality improvement project or policy. This can be particularly useful in identifying facilitators or barriers to the development of a research project or its implementation, the development of a health care service or implementation of a policy.

How to conduct a stakeholder analysis. A *stakeholder analysis* can be a useful tool when introducing a new policy or implementing policy recommendations as well as when embarking on changes to the health care delivery service, evaluating an initiative or other organizational changes. Getting the opinions of those that will be affected by the change is important to facilitate the success of the planned change/s. A *stakeholder analysis* is also helpful in identifying opportunities and threats to the proposed changes, which will assist with decision-making. Undertaking a *stakeholder analysis* involves a systematic approach utilizing a similar methodology to that of a research project. It will include aims/objectives, methodology, data collection, data analysis, discussion of findings in relation to existing data and dissemination as well as any limitations and/or implications.

When conducting a *stakeholder analysis* there are a series of steps to be followed beginning with questions to be considered before the analysis begins [2]. These include the context and scope (individual, organizational or national) of the analysis, what is its purpose, what are the aims and objectives of the analysis what methodology will be used to undertake the *stakeholder analysis*, how will the data be collected and analyzed, who will undertake the data analysis – a team or an individual? Consideration must also be given to the time frame (short or long) and this may depend on the budget available for the stakeholder analysis as well as the purpose. Other factors to be taken into account are how the findings will be presented, how any proposed change will be implemented and sustained and how any limitations of the *stakeholder analysis* will be presented. It is worth noting that conflicts can arise and that not all members of a team or organisation will favor a *stakeholder analysis* so strategies to manage such behaviors also need to be in place.

Changing philosophy in the delivery of healthcare. Across many health care systems the underpinning philosophy is changing from what was a traditional medical model to a more social approach. At organisational, team and individual levels this can present challenges as it calls for a change in culture and clinical practice. As part of this culture change greater emphasis is now given to person-centred-care [5], which fosters therapeutic relationships between clinicians, individuals and their significant others underpinned by values of mutual respect and individual right to self-determination. Shared decision-making is an essential part of this process and honours and values the voices of those with health care problems. It is predicated on enabling individuals to speak up during a clinical consultation as opposed to isolating people in their experience of suffering and resilience [6]. Enabling patients, carers and service users to contribute in this way increases their sense of self-worth, self-esteem and offers them a degree of ownership.

Stakeholder analysis of user involvement in resilient healthcare. This section will consider stakeholder analysis of user involvement in *resilient healthcare* with emphasis on mental health, but the approach is equally applicable to other areas of healthcare. There is a growing awareness that mental illness has a bearing on all aspects of society from individual experiences to wider economic impact. With this increasing awareness is the acknowledgement that those experiencing mental health issues should have greater input into the decision-making process about their care including service development although this is not always systematically conducted. Engaging service users in this way requires professional competencies and the communication skills that encourages service users to have a more active role in their own treatment and care [7]. Reference [8] highlight that patients, family and healthcare stakeholders are fundamental co-creators of resilience and the introduction of recovery oriented, person centered care provides such an opportunity. However, [9] indicated that inpatients reported few opportunities to have meaningful input into decision-making about their care. From the providers' *(stakeholders)* perspective, patients' were perceived as difficult to engage in care planning, goal setting and in meetings about treatment. Furthermore, [7] suggest that for service transformation to take place, providers need to understand and experience working with those undergoing mental health issues with mental health disorders in different roles and positions hence the importance of *stakeholder analysis* and the promotion of *resilient healthcare*.

To achieve greater service user involvement and greater integration we suggest a modified *stakeholder analysis* drawing on the approach of [2] for the organizational level and for the planning of individual care the work of [3]. Integrating these two approaches into the *stakeholder analysis* will ensure a stronger working relationship between the organisation and those it offers care to. Having the service users participate in this way will ensure a better match between the aspirations of both the individual and the organisation resulting in the *stakeholder analysis* being co-produced [10, 11]. Using a co-production approach enables service users/ patients to work in partnership with researchers, academics and/or clinicians. This partnership working can take a variety of forms including service users as researchers, or as members of advisory or steering groups. Prior to commencing a *stakeholder analysis* at the organizational level the benefits of such an exercise should be considered. Theoretically, the outcome should be that it will lead to enhanced service quality including better quality care, enhanced care outcomes, improved working relationships between practitioners and service users and overall improved care outcomes for the organisation.

As highlighted earlier when using the approach of [2] to conduct a *stakeholder analysis* a number of questions need to be addressed such as what is the context? This will depend on both the historical and contemporary culture of the organisation and its ethical principles. At what level with the analysis take place? This can take place at all organizational levels or more widely depending on the key questions as to why a *stakeholder analysis* is required. In order to conduct the analysis the stakeholders need to be identified and the best way to recruit them as well as consent to participate bearing in mind that not everyone maybe willing to take part. The nature and type of data that is required and the methods of its collection need to be agreed. Who will conduct the data analysis is a further consideration and can be either at team or individual level depending on the nature and objectives of the *stakeholder analysis*. The *stakeholder analysis* can be conducted by those internal to the organisation but also external, bearing in mind there are *pros* and *cons* to both. What will be the time line for the *stakeholder analysis*? For healthcare organisations this can be within a twelve month period or related to financial year returns. Other considerations are the reliability and validity of the data and any limitations of the *stakeholder analysis*; how the findings will be presented and by who needs to be addressed including the role of the service user participants as sometimes their contribution can be overlooked? Further important considerations of the *stakeholder analysis* are how the findings will be evaluated, used and any suggested changes implemented but this will depend on the objectives of the analysis.

Moving from *stakeholder analysis* at organisational level to the analysis at an individual level where the work of [3] will be considered as a framework. This framework highlights how individuals can be involved at different levels in the decision making process from their own treatment decisions to that of policy making and can act as a guide to individual involvement in *stakeholder analysis* making it more relevant to the lay public as they have a stake in the process. Within the framework there are three variables: (1) Decision-making domains (2) Role perspectives and (3) Level of participation. The decision making domains refer to treatment, service delivery and broad macro-or system-level decision-making. These subdomains are not entirely independent, for example the domain referring to treatment decisions takes account of the treatments/interventions that are available to patients whilst the second domain relates to resource allocation and the services that can be accessed and by whom in the defined locality. The third subdomain relates more to broader health care allocation and policy decisions at wider national levels.

The second variable in the framework is role perspectives where individuals can take on a variety of roles in the decision-making process including patient, advocate, peer-support worker, volunteer or policy maker. It is now better recognized that individuals can bring different perspectives to the health care decision-making process such as their role as a service user and a public policy perspective. Having the service user perspective can highlight concerns or benefits of any care decisions on health at an individual level, interest or support groups and the wider community. A public policy perspective takes on a much wider view and reflects a concern for the wider public good rather than specific more personal interests. The distinction is important because each role perspective incorporates different attitudinal assumptions and expectations, which individuals then bring to a particular decision-making context.

Focusing on the level of participation and the extent to which individuals have control over the decision-making process is the purpose of the third variable. In the literature there are a variety of terms used to describe the level of participation from consultation to service user control or service user led. Reference [3] state that in order to keep their framework manageable but at the same time capturing important distinctions in decision-making control, they collapsed their ladder into three categories: consultation, partnership, and lay control. These distinctions indicate the different level of patient participation in health care decision making. The framework acts as an analytic tool for conceptualizing key dimensions of lay involvement in health care decision-making. It provides a systematic structure for classifying a range of options available for lay participation in healthcare decision making and a useful template when conducting a *stakeholder analysis* around service user involvement in mental healthcare. Table 14.1 below offers a summary of the preparation and process for undertaking a modified *stakeholder analysis*.

Table 14.1 Modified stakeholder analysis: preparation and process

Preparatory questions

- What is the purpose; aim and objectives of the analysis
- What is the context and scope
- Who will be the stakeholders
- What will be the timeframe
- Is there an identified budget for the analysis
- How will patients and public be involved
- Who will undertake the stakeholder analysis - will it be internal or external to the organisation
- What will be the methodological approach
- What methods of data collection will be used and who will collect the data
- How will the findings be presented to the organisation and by whom

Process

- Recruitment and engagement of the stakeholders
- Data collection and analysis
- Drafting and agreeing on the stakeholder analysis report
- Reporting the findings
- Planning for implementation of any changes that emerge from the analysis
- Planning for sustainability

14.3   Results of a Stakeholder Analysis

With respect to the movement towards person-centred care and shared decision-making conducting a *stakeholder analysis* will give senior managers good insight into the opportunities and barriers towards change. It will highlight the organisations readiness for change and any education or managerial changes that need to take place and the pace at which the change should be introduced. Additionally, it will indicate the level of service user involvement in care decision-making, in all aspects of organisational activity and wider policy development. The overall outcome will lead to enhanced service quality including better quality care, enhanced care outcomes, improved working relationships between practitioners and service users and overall improved care outcomes for the organisation.

14.4   Conclusion

This chapter has explored the application of a modified *stakeholder analysis* in health-care together with a conceptual framework focusing on lay participation in healthcare decision making. The importance of planning for a *stakeholder analysis* was indicated as well as the key questions that need to be considered beforehand including context and methodology. The role that patients and public can play in a *stakeholder analysis* was also considered along with benefits and potential results.

References

1. J. Braithwaite, R.L. Wears, E. Hollnagel, Resilient health care: turning patient safety on its head. Int. J. Qual. Heal. Care **27**(5), 418–420 (2015)
2. R. Brugha, Z. Varvasovszky, Stakeholder analysis: a review. Heal. Policy Plan. **15**(3), 239–246 (2000)
3. C. Charles, S. DeMaio, Lay participation in health care decision making: a conceptual framework. J. Heal. Policy Law **18**(4), 881–904 (1993)
4. M.E. Clarkson, A stakeholder framework for analysing and evaluating corporate social performance. Acad. Manag. Rev. **20**(1), 92–117 (1995)
5. B. McCormack, T. McCance, *Person-Centred Nursing: Theory and Practice* (Wiley-Blackwell, 2010)
6. R.E. Drake, P.E. Deggan, C. Rapp, The promise of shared decision making in mental health. Psychiatr. Rehabil. J. **34**(1), 7–13 (2010)
7. M. Storm, A. Edwards, Models of user involvement in the mental health context: intentions and implementation challenges. Psychiatr. Q. **84**(3), 313–327 (2013)
8. C.C. Schubert, R. Wears, R.J. Holden, G.S. Hunte, Patients as a source of resilience, in *Resilient Health Care, Volume 2: The Resilience of Everyday Clinical Work*, ed. by R.L. Wears, E. Hollnagel, J. Braithwaite (Ashgate, 2015), pp. 207–225
9. M. Storm, K. Hausken, K. Knudsen, Inpatient service providers' perspectives on service user involvement in Norwegian community mental health centers. Int. J. Soc. Psychiatry **57**(6), 551–563 (2011)
10. S. Gillard, R. Borschmann, K. Turner, N. Goodrich-Purnell, K. Lovell, M. Chambers, Producing different analytical narratives, coproducing integrated analytical narrative: a qualitative study of UK detained mental health patient experience involving service user researchers. Int. J. Soc. Res. Methodol. **15**(3), 239–254 (2012)
11. M. Chambers, Engaging patients and public in decision-making: approaches to achieving this in a complex environment. Heal. Expect. **20**(2), 185–187 (2017)

Chapter 15
Resilience: From Practice to Theory and Back Again

Carl Macrae and Siri Wiig

Abstract This book offers a purposefully broad exploration of resilience: it presents a variety of diverse perspectives in a range of practical contexts across various scales of system from a range of disciplinary positions. One of the core organising principles of this book is a concern with understanding how ideas of resilience can be translated into practice, and how practices of resilience can in turn be theorised and explained—irrespective of whether those practices are conducted at the 'street-level' by frontline actors or in the committee rooms of policymakers. To do this, the book explores empirical, methodological and theoretical challenges in analysing resilience, defining resilience, organising resilience, building resilience, leading resilience and regulating resilience—to name just a few of the activities that provide the focus of concern in these chapters. In this chapter, we provide a brief and necessarily partial survey of the varieties and commonalities of resilience that have emerged throughout the book, and then explore how—and why—we might move towards an integrated theoretical framework of resilience.

Keywords Theory · Integrative framework · Resilience studies · Systems improvement

15.1 Varieties of Resilience: A Tour of the Landscape

Throughout this book one of the most apparent, and perhaps inevitable, characteristics of resilience is its variety. From the basic definitions and objectives of resilience to the ways in which it is operationalised in different settings, it is clear that one of the strengths of the concept of resilience is its ability to accommodate a broad set

C. Macrae (✉)
University of Nottingham, Nottingham, UK
e-mail: carlmacrae@mac.com

S. Wiig
Faculty of Health Sciences, SHARE-Center for Resilience in Healthcare,
University of Stavanger, Stavanger, Norway

© The Author(s) 2019
S. Wiig and B. Fahlbruch (eds.), *Exploring Resilience*, SpringerBriefs
in Safety Management, https://doi.org/10.1007/978-3-030-03189-3_15

of phenomena in many settings and at different levels and scales of activity. This allows disparate and diverse practices to be analysed through the same conceptual lens—from the second-by-second work of air traffic controllers (Woltjer, Chap. 13) to the long-term organisation of global humanitarian efforts (Kimber, Chap. 11; Tjoflat and Hansen, Chap. 12). However, this breadth can also be a source of weakness, and points to the risk of theoretical over-reach: if a concept attempts to explain everything it can end up explaining nothing. Nonetheless, across the book a set of core ideas and principles can be seen emerging amongst the varied perspectives. For instance, definitions of what resilience *is* clearly vary, but there is general agreement on core characteristics: primarily the ability of an entity—individuals, communities, organisational units or larger systems—to return to some 'normal' condition or state of functioning after an event that disrupts its state; or to adapt to a new normal state, where system functioning is reorganised or enhanced in some way in response to the disruption. Basic definitional issues such as this are not simply of theoretical concern but have deeply practical consequences. As Kimber (Chap. 11) argues, a vague definition of resilience presents challenges for creating and managing resilience in operational activities—while also acting as a valuable 'boundary object' that can bring together different communities around a (loosely defined) shared goal. In practice, as in theory, the emerging field of resilience studies needs to engage in an ongoing negotiation to ensure theories of resilience can both support the benefits of coordinating diverse communities and perspectives, while also providing the degree of detail and specificity needed to operationalise and use those theories in different domains.

Another variety of resilience explored throughout the book concerns the level of analysis and scale of activity that provides the focus for understanding resilience. Resilience is inherently a systems-oriented concept: it can be applied at different levels of a system, from the level of individual cognition to entire societies and beyond; and it can be used to examine the complex interrelations and interconnections between different levels and scales of these systems. Again, this breadth would appear to be both a source of strength and weakness. The moment-by-moment activities and cognitive processes of individuals can seem far removed from the large-scale adaptations and reorganisations that unfold across entire industries, and studying these interrelationships at vastly different scales of time and space presents considerable methodological challenges. As Le Coze describes (Chap. 2), these challenges include long time scales, complex networks of stakeholders, distributed research sites, and a large quantity and wide variety of data. Nonetheless, the work presented in this book indicates that, to better grapple with key issues in theorising and operationalising resilience, we need to be much more attentive to the levels or 'scales' of activity at which resilience unfolds, along with the different forms, functions and characteristics of resilience at these different scales of activity. Kyriakidis and Dang (Chap. 6) and Woltjer (Chap. 13) use the domains of critical infrastructure and air traffic control to explore how resilience depends on interactions and adaptations that interconnect across different system levels, and analyse how people at all levels of a system can contribute (or not) to resilience by adapting performance to local conditions. For example, Kyriakidis and Dang (Chap. 6) argue that frontline personnel are involved in short-term adaptations when monitoring and responding to service deviation, but it requires higher-order managerial and organisational responses to effectively learn from an organizations' past experience and anticipate future threats—processes they conceptualise as a 'resilience capability loop' that organises the activities of anticipation, monitoring, responding, and learning. Some of these ideas are echoed in Macrae's (Chap. 3) framework of resilience at different scales of sociotechnical activity, from situated and immediate responses that unfold rapidly, to structural adaptations that involve longer processes of reorganisation, to long-term systemic reconfigurations involving system-wide reform.

One of the challenges of understanding resilience as a truly system-level phenomenon across different scales of activity is exemplified by the work of Berg and Aase (Chap. 10) in the context of healthcare: research on resilience has, to date, mostly focused on those working at the 'front line' or 'sharp end' of organisational practice. And, in the domain of healthcare, little systematic empirical attention has been paid to the higher-level systems activities at the macro level, encompassing the role of national bodies and regulatory agencies. This knowledge gap echoes experience in other fields, where our understanding of resilience mechanisms within and across regulatory actors, or the networks that span entire industries, is lacking (eg Woltjer, Chap. 13). For example, what forms does ongoing adaptation take within regulators, in response to challenge and feedback from regulated organizations, the public, or stakeholders [1]? More broadly, understanding the roles of different stakeholders in the active 'co-creation' of resilience is emerging as a key issue and focus for future empirical work, particularly in the contexts of healthcare, disaster planning and recovery, and international operations. Tjoflåt and Hansen's (Chap. 12) experience with humanitarian programs in protracted conflicts indicate how working in partnership with people and local communities is essential in building resilience. Similarly, Baram (Chap. 5) describes how acknowledging and valuing inter-organizational dependencies is central to building resilience in public water supply systems. Locally-based organizations and networks of other stakeholders can provide public support and contribute to both preventative and recovery efforts in critical infrastructure. One of the practical implications of this is explored by Chambers and Storm (Chap. 14). Understanding the co-creation of resilience requires more systematic and routine stakeholder analysis—both in research and practice—to identify and maximise mechanisms of co-creating resilience across levels of socio-technical systems, and also to assess potential 'risk makers' who may destabilise those efforts of co-creation.

Ultimately, perhaps one of the most practical issues explored by the book is: to what extent is it possible to design and implement resilience—as an individual, an organisation, a leader, or across an entire system? And, what strategies or methods might support practitioners in this? Leadership is particularly important in creating resilient organisations, and some of the key leadership challenges are described by Grote (Chap. 8)—particularly the ongoing and delicate balance required to match stability and flexibility demands in organizations and create environments for resilient performance. Leaders themselves, just like their organisations, need to be able to function in different modes of operation, train their adaptive capabilities and manage rapid mode shifts in response to changing conditions. Sophisticated organisational structures and processes are also needed to support resilience. Pettersen Gould (Chap. 7) examines how airlines adopt meso-level strategies of resilience that support the collection of organisation-wide operational data, operational planning and design, and coordinated action to adapt and improve operational activities. By promoting continuous monitoring, analysis, and attentiveness to possible adaptions in pilots' landing approaches, airline managers enact systems that support 'precursor resilience'—rapidly addressing disruptions and fluctuations to maintain safe performance. At the level of specific organisational interventions, both Anderson et al. (Chap. 4) and Reiman and Viitanen (Chap. 9) provide rich demonstrations of how particular improvement methods, checklists and related sociotechnical tools can be implemented to support adaptive processes of improvement and safety management. Drawing on the contrasting settings of healthcare, nuclear power and beyond, Anderson et al., Reiman and Viitanen document ways in which a variety of practical tools can be integrated into (and evaluated as part of) more strategic programmes of resilience engineering.

15.2   Towards an Integrative Framework of Resilience

Taken together, these chapters sketch out a rich variety of applications, complications and opportunities for the field of resilience, in both theory and practice. This variety points to the wide landscape of resilience that is only beginning to be systematically explored. This variety also presents clear challenges for the future of this developing field. These include the challenges of communicating across disciplinary perspectives, working at different scales of system and levels of analysis, integrating insights across different applied domains and practical settings, and ultimately connecting theory and practice in ways that are productive for both. It is also striking that many of the challenges of developing this field echo those of enacting resilience itself—such as engaging diverse stakeholders, adaptively shifting perspectives and modes of operation depending on context, and updating and integrating models to account for local variation. Addressing these challenges points to the potential value of a broad, integrative framework to help support and coordinate cross-domain, international, multi-level and interdisciplinary work on understanding and operationalizing resilience in complex sociotechnical systems. Resilience is an expansive field of research encompassing many different domains and issues—indeed, that is one of its defining strengths. Clearly no single theory could be expected to properly capture and explain all the relevant factors, concepts, relations, and logics that would constitute a fully developed and all-encompassing theory [2]. However, mapping out the general contours of a broad and expansive framework may act as a useful coordinating device for a range of future work, and help to highlight future questions and shape future objectives in both the research community and for practitioners in high-risk industries.

Building on the work in this book, one useful place to start in developing such a framework would be to identify the underlying and most basic commitments that might underpin an integrative framework of resilience. Perhaps the most foundational principle here is that all organised human and technical activity is constituted by some degree of inherent fluctuation and variation; and that resilience represents the active and effortful application of different sociotechnical resources (skills, knowledge, relationships, equipment, values, creativity, etc) to handle those moments of disruption that threaten current goals. This basic principle appears to rest on four core assumptions that guide much work in this field, and this book. First, ideas of resilience focus on, and have a deep concern with, acting in the world; a key focus of resilience is always some sort of ongoing practical, situated activity in a complex, socially organized 'real-world' setting. A second assumption is that mechanisms of resilience broadly involve the deployment of certain skills, capabilities or resources to handle a particular challenge or stress. A third assumption is that processes of resilience are inherently dynamic, involving change and adaptation in complex multifactorial and multi-level systems. A forth assumption is that resilience, in its broadest conception, is fundamentally a dualistic concept: it is about conserving and maintaining some set of functions or goals, and achieving this through (or despite) changes, adaptations and reformulations of ongoing activities and performances.

While this list is by no means exhaustive, this core principle and these four assumptions appear to exemplify commonalities across many ideas and ideals of resilience, and appear to be woven through many of the underlying commitments that are explored and operationalised throughout this book. If this is correct, then these are the types of basic commitments that can help form the basis of a future integrative framework of resilience. They also point to some of the more granular questions that can frame an integrated and programmatic inquiry into the nature of resilience, and that help to transcend any particular disciplinary traditions or domains of application. These questions include:

- What are the core *phenomena* of resilience? Where and in what ways does resilience unfold in a sociotechnical system, and what is the core focus for study and for management?
- What is the *context* of resilience? Which sociotechnical factors, processes and elements are central to take into account when explaining and engineering resilience, and which are merely peripheral?
- What are the types of sociotechnical *resources* that are drawn on to support and enact resilience? How are these drawn on, applied, used, created and renewed? What counts as a resource, to whom, in what circumstances and for what purposes?
- What are the *processes* through which resilience unfolds? What are the sociotechnical mechanisms that support ongoing adaptation and adjustment? And how are these processes organised, designed and managed?

Many of the explorations in this book, and elsewhere in the literature, offer useful perspectives on these questions. Indeed, any discussion of resilience or effort to operationalize resilience must be based on answers to these questions. However, these answers can often be implicitly assumed rather than explicitly defined. These questions therefore represent another jumping-off point for developing an integrative framework for resilience, helping to define its shape and content. To develop this

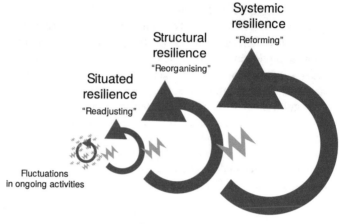

Fig. 15.1 Resilience at three scales of sociotechnical activity

further, it is useful to revisit the analysis presented by Macrae in Chap. 3 that seeks to characterise resilience at three different scales of sociotechnical activity: situated, structural and systemic. The core assumption is that all organisational activity is inherently variable, and that at some point, those variations can become disruptions to ongoing activities and require the *situated* application of existing sociotechnical resources to respond, readjust and recover ongoing activities. Sometimes, existing resources may not be adequate to address the scale of disruption at hand, and so structural adaptations may be required, to reorganise existing sociotechnical resources and associated situated practices. On occasion, the arrangements for enacting structural adaptations may not be adequate to address the scale of disruption that has emerged, and so systemic adaptations may be necessary, to reform the systems, processes and structures through which sociotechnical resources and situated practices are designed and produced (Fig. 15.1).

This framework seeks to move beyond definitions of resilience that are tied to particular 'levels' of system structure, or 'types' of resilient capability. Instead, the focus is primarily on providing a general framework and initial language for explaining the processes of resilience at different scales of time and space in complex sociotechnical systems. For instance, activities in what are traditionally viewed as 'macro' level settings (such as national regulatory agencies or supranational bodies) may still represent processes of situated resilience: small scale adaptations to disruptions by applying available and pre-existing sociotechnical resources. Conceptualising resilience in terms of situated practices and sociotechnical resources can help distinguish between activities of resilience that are fundamentally *systemic*—that involve deeply reforming the foundational elements of how a sociotechnical system is organized and constituted—as opposed to activities that are simply *widespread* but involve no change to the fundamental model of the system, such as lots of people updating a simple local protocol in many different settings across an industry.

Framing resilience in this way provides one example of how an integrative framework might build on the core principles, assumptions and questions outlined above, to move beyond categories and concepts that can sometimes inadvertently maintain disciplinary boundaries or introduce artificial silos into the analysis and explanation of resilient systems. For example, this framing of situated, structural and systemic resilience can provide a framework for focusing attention on the critical questions of how, when and why fluctuations become disruptions, and how disruptions expand in scale to provoke larger moments of resilience that increasingly enroll greater numbers of stakeholders across a system and ultimately challenge, reorganize and reform core elements of that system. To take one practical example from healthcare: media reports and public pressure relating to the handling of a harmful adverse event can act as a significant disruption for a regulatory body, prompting a rapid response to review and reopen an investigation using existing organisational resources and models (situated resilience); which in turn can provoke the design and reorganization of new processes for involving next-of-kin and outside experts in inquiries (structural resilience); and ultimately can lead to system-wide recommendations for reforming the underling mechanisms of collaboration between regulatory bodies involved in assessing compliance with core regulatory requirements (systemic resilience) [3]. This example illustrates how an integrative framework for resilience should be flexible enough to accommodate granular details related to a single organization and event, whilst also providing a language to explain the 'scaling up' of resilience across entire systems and over long time periods. An integrative framework should also accommodate the 'positive', as well as the 'negative', aspects of resilience: the processes of improvement, adaptation and innovation as much as the management of the adverse impacts and crises that are often viewed as the prime triggers of resilience. For instance, in the safety sciences the language of 'disruption' may seem inherently skewed towards the dark side of organisational life and suggestive of harmful or adverse events. However, when viewed through the lens of a framework that is as much focused on the 'positive' processes of adaptation, change and renewal, then 'disruption' can take on a broader meaning. The situations that provoke resilience might always be defined as 'disruptive' to an existing way of doing things or to established cultural assumptions—but that disruption might be in the form of a potentially harmful event, or in the form of the invention of a new and innovative technology. Both of these represent a disruption to an existing world order. An integrative framework of resilience should be able to fully accommodate, and explain, both.

15.3 Resilience: Between Theory and Practice

Resilience is a powerful and engaging idea because of its breadth and scope. Likewise, its breadth and scope leave the field of resilience studies open to increasing fragmentation and polarization. Accordingly, it seems important that the field moves towards debates that surface the underlying assumptions, principles, commitments and questions that support cross-pollination of ideas, insights and practices. Developing an integrated framework that can accommodate a wide range of concepts, strategies and models of resilience, and offers a way of connecting them and linking them across levels of activity, seems an important part of this journey. In the words of Kurt Lewin, there is nothing quite so practical as a good theory [4]. A good theory should be able to explain, predict, and delight [5]. And theories should be practical things that help us act in, and on, the world. Due to the many varieties of resilience, a single, simple and general theory of resilience is not a realistic objective. But this book indicates that the field would benefit from engaging in a thorough, expansive and ongoing process of theorizing [6] that continues to define and address the many cultural, symbolic, organisational, sociotechnical and practical aspects of resilience that operate across myriad scales of time and space.

References

1. S.H. Berg, K. Akerjordet, M. Ekstedt, K. Aase, Methodological strategies in resilient health care studies: An integrative review (2018). Resubmitted to Safety Science, referred to with permission from authors
2. D.A. Whetten, What constitutes a theoretical contribution? Acad. Manag. Rev. **14**(4), 490–495 (1989)
3. S. Wiig, K. Aase, M. Bourrier, O. Rise, Transparency in health care – displaying adverse events to the public, in *Risk Communication for the Future*, ed. by M. Bourrier, C. Bieder (Springer, Berlin, 2018), pp. 111–125
4. A.H.V. de Ven, Nothing is quite so practical as a good theory. Acad. Manag. Rev. **14**(4), 486–489 (1989)
5. R.I. Sutton, B.M. Staw, What theory is not. Adm. Sci. Q. **40**(3), 371–384 (1995)
6. K.E. Weick, What theory is not, theorizing is. Adm. Sci. Q. **40**(3), 385–390 (1995)

Printed in the United States
By Bookmasters